FORMATION AND USE OF
INDUSTRIAL BY-PRODUCTS:
A GUIDE

Frontispiece: Sludge gas engines at Katherines Farm Sewage Works.
Courtesy: City and County of Bristol and W.H. Allen Sons & Company
Limited

FORMATION AND USE OF INDUSTRIAL BY-PRODUCTS: A GUIDE

by A.W. NEAL, CGIA, CEng, FIMechE, FIWE

BUSINESS BOOKS LIMITED
London

ISBN 0 220 66255 x

By the same author
INDUSTRIAL WASTE

Prepared for press by the Ivory Head Press,
30 Craven Street, London WC2
and printed in England by
Redwood, Burn Limited, Trowbridge and Esher
for the publishers, Business Books Limited
(registered office: 110 Fleet Street, London EC4)
publishing office: Mercury House, Waterloo Road, London SE1

CONTENTS

INTRODUCTION

The favourable manner in which my book on Industrial Waste has been received has lead me to develop this book on an allied subject of equal interest to those concerned with the further use of materials.

The object of the present work is the same as the last — mainly to furnish practical information in non-specialist language. Some unfamiliar words and phrases are inevitable; a short glossary has been provided to explain such words and phrases (see end of Chapter 1).

To rouse interest I have tackled the subject on a broad front and in general terms. If my reader has the appetite of an Oliver Twist, I am comforted that there are specialised publications to which he may refer, some of which are mentioned in the text.

I owe a debt of gratitude to friends who have provided advice and material, and I owe a special debt to authors too numerous to mention individually. I am also grateful for the help afforded by the staff of my publishers who navigated the manuscript through the tangles of production.

Looking at the finished product, I am conscious of omissions, but what has been left out is largely from sheer necessity. To attempt to include more would have been an unwieldy task and would have made this book inordinately long and I thought it far better not to overload.

A.W. NEAL

Chapter 1

GENERAL REVIEW

The environment in which we live is becoming increasingly polluted as a result of the indiscriminate discharge of poisons and ecologically undesirable by-products. The sweeping technological advances of the last few decades have in many cases been made with little care and attention to the results of such pollution so that we have reached the situation today where there is too much pollution, too much waste, too much waste, too much noise and too little cleansing. One might say that we are now reaping the more unpleasant rewards of the Industrial Revolution.

If the blame for pollution is laid at the doors of industry, then it is for industry to provide the cure. Optimists admit that there are problems, but think that there is plenty of time to solve them. One large problem is economic in character since pollution is rooted in waste, and the disposal or alternative use of waste is tied up with economic feasibility.

Besides pollution, another critical situation is developing fast — the internationally based 'energy crisis'. Forecasters envisage an ever-widening gap between supply and demand for fossil fuels and something must be done urgently if we are not to leave our descendents a world that has no power.

These serious uncertainties are not confined to fuels. They are making themselves felt in every field where things are made or prepared and distributed for marketing, especially in industries that are dependent on foreign sources of raw materials.

Having said this, what are the main fronts on which these problems should be tackled? Dispassionate observers say there are three. First, in the short term, production must be increased; secondly, in the long term, substitutes must be found; finally, materials must be conserved by end-use efficiency.

Both matters — pollution and supply/demand — benefit by better over-all use of raw materials in the manufacturing stages. This signifies the need for reclamation and re-use of valuable fractions — in many industries a routine practice. There seems, however, to be some areas in which the maximum end-use efficiency has not yet been reached. How far such operations could be extended is a matter for those concerned, but the limit should not be set by financial considerations alone. Waste entails eventual disposal — a growing worry with dumping sites shrinking and anti-pollution legislation growing. As it is local authorities now shoulder most of the load, but it can not be expected, or foreseen, that they will process valuables other than those contained in assorted refuse which it is their duty to receive.

By-products are the essence of the following chapters. Their separation from the main product may be simple or complicated. It is a business that has grown beyond all expectations, and is still growing. Some branches are extensions of traditional practices, whilst others are entirely new.

The term 'by-product' is loosely applied everywhere and difficult to pin down to a precise meaning. *The Comprehensive Dictionary*, published in 1863, does not even mention it, presumably because it was not invented at that time. One popular dictionary's interpretation gives a by-product as 'something evolved in a process, especially a manufacturing process, apart from the chief result.' An explanation given by a well known dictionary of science is 'a substance, useful or otherwise, obtained incidently during manufacturing some other substance, often as important as the manufactured substance itself.' Both of these interpretations are correct in so far as everyday usage is concerned, but in industrial parlance the prefix does not always denote something subsidiary. In part, the difficulty of definition lies in the fact that the different branches of industry have, by custom, their own distinctions for by-products. This is certainly true in the multi-track petroleum industry. The term by-product hardly ever appears in the oil industry's own vocabulary, and it does not read happily to the oilman. It can be argued that a refinery constructed to produce motor spirit means that every other product is consequently a by-product. Conversely the main interest may be fuel oil and so everything else would be a by-product. In fact an oil refinery nowadays is not purposely directed at motor spirit alone. It is designed to produce a very wide range of commodities. So the industry divides the many derivatives under three headings: Main Products, Products and Specialities.

An anomaly occurs in the meat preparation trade where carcases of slaughtered animals are separated into two main groups — the Products (edible parts) and the By-products (inedible parts). Neither have been produced by a manufacturing process. A similar deviation arises in the milk processing business.

The three main items are Butter, Cream and Cheese. But the processing does not rest there as there has been a whole variety of dairy products evolved from these. This trade refers to these additional preparations as by-products, although none are obtained accidently.

In the case of the sugar beet business the dictionary explanation is precise. The sugar is processed from the plant and the residuum becomes a feed for animals.

This matter of when is a by-product not a by-product arises again and again, sometimes in quite unexpected ways. One example is associated with the annual culling of seal pups. The object behind this non-commercial scheme is to keep the seal population in this country within reasonable limits. The number of allowable slaughterings is fixed after the total number has been estimated. This total, arrived at by tagging, is put forward by the Government-sponsored National Research Council. The Universities Federation for Animal Welfare, however, say that its own survey, carried out by helicopter, gives a much lower but more accurate figure. If this is correct then the seals would eventually be wiped out altogether. Why cannot something be agreed upon? The Justice for Animals Society say 'The killings are purely commercial and the Home Office should not give licences to kill to anyone'. The pups, when about ten days old, are coveted for their soft skins. When slaughtered their bodies go for fertiliser and animal feeding stuffs. Perfect skins, for which the hunter is paid about £40 each, are used mainly for making fur coats and handbags. These are the commercial by-products from an otherwise non-commercial activity.

Employed here and there in this book is the term 'collateral products'. The dictionary explanations of the word 'collateral' are in agreement, although some people will contend that it refers to persons rather than things. Collateral means 'common decent from the same stock', but a different line, or subordinate from the same stock. This seems to be a handy term to describe products that do not strictly fall under the general heading of by-products. The borax family of products would fall under this heading.

At the bottom of the manufacturing scale are residual substances that are disposed of to anyone willing to afford them accommodation, or just dumped — thus occasioning a potential danger. The policy of past and present Governments has been to enforce anti-pollution legislation with great vigour where waste is hazardous. Disposal by dumping is becoming more difficult and both it and the cost of treatment is multiplying accordingly. With rising prices many wastes must be regarded as by-products whose sale can offset the cost of treatment.

At the high end of the scale are the marketable by-products which are as profit-yielding as the principal lines. Nevertheless, these are expected to produce some useless residues.

Important points to note are:

1   Britain imports 90 per cent of her raw materials and 50 per cent of her food (the over-all EEC figures are, one imagines, similar).

2   Britain is short of foreign currency.

3   As the countries from which Britain imports food and raw materials develop their own industries, sales to them of our manufactured goods will fall and our imports from them will consequently become dearer.

4   Many raw materials, as they become scarcer, will be regarded by the source countries as strategic materials that they do not wish to sell at any price.

The need for more emphasis on waste and material recovery should be manifest from these points.

Although there is disparity in the degree of progress made between individual industries, and bearing in mind that there is a limit to what can be done, recent technical advances have given a much needed boost to the commercial exploitation of waste. Now that most branches of industry have their own research organisations there is far more free-flowing information than hitherto. To this should be added the recourse manufacturers now have to Government research establishments. Further, there are the independent consultants who are prepared to assess values, to seek fresh approaches to old problems and to enquire into new ones. The specialists are also ready to examine reports by others and to advise whether their recommendations should be accepted or rejected.

The Government takes the view that all opportunities for efficient and economic use of resources should be closely considered, but it believes this is the prime responsibility of the industries concerned. Where the exploitation of by-products is not economic for these industries the Government's concern is that residual materials are disposed safely and efficiently and legislation is adequate to ensure that this is carried out. As fresh disposal problems arise the law will be extended to include them.

A brief reference should be made to at least two Government aids to industry: the Torry Research Station and the Warren Spring Laboratory. The former, and its out-station at Hull, are laboratories concerned with the investigation of all aspects of fish technology. The establishment possesses a research trawler and mobile laboratories, and staff including engineers, physicists, biochemists and bacteriologists. Advice and assistance in all branches of fish technology including, for example, product development and design of equipment is available. Contract work is undertaken for individual firms and organisations both home and abroad.

Warren Springs is a laboratory for industrial and environmental research whose services are available on a contract basis to industry and the professions, government departments and local authorities. Industrial research, which occupies two-thirds of the research effort, is devoted to mineral science and technology, bulk materials handling, control engineering and catalysis. The laboratory's environmental interests are concerned with air pollution, oil pollution of the sea and beaches, and recovery of valuable materials from scrap and waste.

Mention should also be made of the National Research Development Corporation, although this is a public concern and not a Government department. This body promotes the adoption by industry of new products and processes invented in government laboratories, universities and elsewhere, and advances money where necessary to bring them to a commercially viable stage. It speeds up technological advances by investing money with industrial firms for the development of their own inventions and projects. It does not itself manufacture or trade, nor does it have its own research or development facilities. It does not receive any grants but is financed by the Secretary of State for Trade and Industry with government loans. The Corporation is required to balance its accounts in the long term and it therefore has to conduct its activities on a sound commercial basis. With a few exceptions, it can deal with all types of invention and project. During the past three years the NRDC has received from industry and allied sources proposals to support development of reclamation and recovery schemes and a number of them are now established commercially.

The most comprehensive information the NRDC publish is contained in the Corporation's Annual Reports. It can be seem from the Report for 1971-72 that the Corporation's activities in the field of by-products were not particularly extensive for that year. Although they are continuing and have slightly expanded, it cannot be claimed that their volume is such as to have a substantial economic result in terms of the nation's trade. It is true that for some time industry does not appear to have mounted long-term development projects on a significant scale. This is, of course, a generalisation. It is believed that the climate is now changing and that the situation will be reversed. The Corporation has been in existence for 25 years, during which the economic climate has varied from time to time. The present low level of activity results from a number of reasons, most

of which have been referred to in the financial press, and
which may change in the course of the next few years.

The US journal, *Patent Licencing Gazette*, for May/June
1973 contains an interesting note headed 'R & D is losing
ground' which points this out:

> 'According to an industrywide survey released recently
> by the McGraw-Hill Economic Department, business is
> reluctant to increase its spending for research and
> development. And much of the money that is going to
> R & D is aimed not at finding new products or processes
> but rather at improving existing ones. The survey indi-
> cates that US business expects to perform $21.2-billion
> worth of R & D this year. Approximately half of this
> involves government-sponsored research. Though the
> total is 5% higher than last year, the gain is more than
> offset by inflation.'

It also reports that R & D expenditure was expected
slip to 2.4 per cent of industrial sales in 1973 compared with
2.5 per cent in 1972 — with a further decline to 2.3 per cent
expected in 1976.

Many companies are shifting their research goals. In the
past, R & D expenditure was largely directed toward new
products and process development. But 44 per cent of the
manufacturers surveyed said that their R & D goal was to
improve existing products. As a result, development of new
products seems to be slowing down. By 1976, the survey
says, only 13 per cent of industrial sales ($134.7 billion)
will be new products, down from the 18 per cent that
industry expected in last year's survey for 1975. Only three
industries — instruments, machinery and electrical machinery
— expected 20 per cent or more of their sales three years
hence to come from new products.

Cooperation between a by-product industry and its
natural parent industry makes for a good combination, both
for production and marketing. Usually, but not always, one
organisation has control of both. But there are many
branches of manufacture that have no affinity, and never
have had, between them. Indeed, the products, the needs and

the methods may be completely unknown to each other.
This appears to be a fruitful area to explore. It has been
suggested by some industrialists in this country that, although
significant developments do occur by the bringing together
of ideas from one environment or industry to another the
chances of any particular transfer or conjunction being
fruitful are very low. This hints that any machinery for
exchanging information between industries with no affinity
would produce little in the way of results in relation to the
amount of effort which would have to be applied. Fortunately this somewhat negative approach is not shared everywhere.
The Nordic countries are to establish a 'waste exchange' at the
Swedish Institute for Water and Air Pollution Research. This
is the proposal of the National Federation of Industries in
Denmark, Finland, Norway and Sweden. This exchange will
receive reports on waste and on residues of raw materials and
finished goods that individual companies want to get rid of.
It will then inform companies capable of utilising the waste
or think they can. The outcome of this forward-thinking idea
will be awaited with interest in this country.

The latest plan on toxic wastes is to make it mandatory
for manufacturers to disclose information on all wastes they
are discharging into British waterways and estuaries. This was
announced late in 1972 when it was revealed in Parliament
that the Department of Environment had accepted almost all
of the Royal Commission on Environment Pollution's
recommendations on cleaning estuaries and coastal waters.
A controversial recommendation that the then Government
had shown it would accept was a scheme to require manufacturers to pay for the discharge of industrial waste. Treatment activities would then have to be regarded by industry
as an integral cost of production. The Under-Secretary of
State at the Department of Environment, said in a written
reply to the House of Commons that the Secretary of State
agreed with the Commissioner's call for urgent administra-
tive action and for new legislation. The main points for
action were as follows:

1   Local authorities to seek information from industry
     about discharges through public sewers and estuaries

or sea.

2 River authorities to assume responsibility for monitoring critical pollutants in industrial estuaries.

3 Planning authorities to consult river authorities before any plans that would increase the pollution of an estuary were allowed to proceed.

4 Industrial and sewage effluent discharges to controlled waters to be brought under full control.

5 All discharges to sewers to be controlled; industry to be charged the full cost of disposal of its wastes;

6 Authorities to be empowered to take samples from private sewers from trade premises.

7 River authorities to apply to the Secretary of State for tidal water orders bringing estuaries at risk under full control.

8 The proposed regional water authorities to control all discharges into rivers, estuaries and coastal waters.

9 The Government to take the lead in international arrangements to publish data on monitoring, etc.

10 Biological monitoring of indicator species of animals and plants to be undertaken in addition to chemical monitoring.

11 Proposed regional water authorities to monitor discharges to estuaries and coastal waters.

12 Sewage authorities to have the right to refuse particular constituents in effluents.

Nothing approaching such stringent and extensive control has been envisaged before.

The matter of the industries concerned bearing the whole of the cost will be challenged by them. Some will point out that industry is already responsible for making district rate payments and that any additional expense should be shared by the ratepayers at large. Others — those that have been operating for a long time — may point out that they have the right by custom to make discharges and that if this right is removed they should receive compensation accordingly.

Whatever the outcome it is clear that carrying out any new statutory duties will be reflected in the final cost of manufactured goods. It suggests a further look at the root of the trouble — what to do with toxic wastes? The problem has been with us for a long time and has in the main been consistently neglected. One of the first tasks should be to collect together all existing information on the use and/or recycling of by-products.

At first, the disposal of the waste products resulting from the old method of manufacturing town gas, i.e. pre-Nationalisation when gas works were scattered haphazardly throughout the country, was deplorable, but by the time the process had become outdated the selfsame wastes had become by-products of equal value to the gas manufactured.

Another early example of large-scale recoupment arose from blast-furnace gases. Up to the end of the nineteenth century these were being expelled direct to the atmosphere. Nobody worried about the environmental effects of such discharges except the few in the immediate locality who had to suffer the discomfort. Then the gases were used, first to actuate heat exchangers to raise the forced-draught temperature, and then to fire steam generators. Ultimately they were used to operate internal combustion engines for blowing purposes. But in defence it should be remembered that this last could not have been performed much earlier as the gas engine, prior to this, only emerged from the embryo stage. Eventually some of the largest gas engines in the world at that time were consuming blast-furnace gas. The engines provided the blast air; the forced draught melted the iron out of the ore; the furnace created more gas; the gas heated the air or was fed to the engines. So the engines and furnaces mutually assisted each other, instead of all the obligations being one sided. The impression of one writer of the time was 'We really appear to be within sight of the time when waste will be unknown'. His prediction is further away than ever. For all that, one wonders why so many production engineers do not follow closely in the wake of the early gas engineers.

Whilst discussing waste effluents it would be appropriate to refer, if only briefly, to those arising on the farm. On every stock-raising farm there is a natural supply of animal manures, both solid and liquid, the right use of which forms an integral part of efficient husbandry. The collection and storage of supplies of this material differs between areas and ranges from simple muck spreading to what may be described as manure irrigation, i.e. self-irrigation, with all that that entails. The subject has been repeated again and again in farming literature and reports, and to these the inquisitive reader is referred.

It appears that no practical method of treating silage effluent has yet been contrived. Disposal is either by soakaway or spreading on fallow or grassland. Some good advice on this subject appears in a publication prepared by the Ministry of Agriculture, Fisheries and Food [1.1].

There are a number of reservations regarding the disposal on land of farm effluents. One now attracting considerable attention is the possible seepage of harmful matter into underground water reserves. Another is the drainage, knowingly or otherwise, of the same into watercourses. Both could result in a breach of the *Water Resources Act, 1963*. Upon these and allied matters the farmer should seek expert advice.

It has been reported that an immense amount of perfectly good food is thrown away annually — one third of the total available for human consumption. This astonishing revelation is hardly believable, especially in view of the high incidence of starvation in backward countries. Yet, only in very favourable circumstances could a useful outlet be found for this material, and few forms of utilisation would justify expensive separation and transportation. This in turn means that it is hardly ever in sufficient quantity in one area to justify collection for more than, say, pig swill.

There is more demand for the so-called 'convenience' foods than ever before, and the trend is always upwards. Apart from variety, their popularity is mostly due to the fact that so many housewives now seek outside careers and therefore have less time to spend in the kitchen. The inclination towards prepared and semi-prepared foods is also assisted by distribution through supermarkets. Although home chores are reduced,

more packing material accumulates. This has been reflected
in the character of the domestic refuse collected by local
authorities.

The rise of the prepared food industry, which has ballooned
in more recent years, has had the effect of concentrating surplus and unneeded fractions of animal and vegetable materials.
In this way the companies so concerned have been able to
tackle the subject of useful by-products on a much broader
front than hitherto, often in their own establishments. The
major leftovers are types of provender and fertilisers, although
others do occur.

Not until quite recently did the public become aware of the
farmer's reliance on foreign sources for animal feeding-stuffs.
This was highlighted by the 1972 dockers' strike, the result of
which was this material becoming strikebound. Farmers depend on the supply of soya and maize from America, fishmeal
from Peru and Scandinavia, and protein, in the form of
groundnut and linseed cake, from India, Burma and South
America. Thousands of tons of cereals and protein were lying
offshore and in barges because of this confrontation. The
Compound Animal Feeding Stuffs Manufacturers Association
was soon asking farmers to help suppliers by not demanding
special 'mixes', but to take what was on offer. At the same
time farmers were giving warnings that they would be in
trouble if the strike continued much longer. Fortunately, a
truce was agreed upon. To the general public there appeared
no good reason why self-sufficiency could not be accomplish-
ed in this field.

Most by-products are accepted today with the same con-
fidence as is placed on the materials from which they originated, and it would be true to say that many of the public
are quite unaware that they are derivatives. But there remains
a lingering belief in some quarters that they might be substitutes. Certainly their collective name does not inspire con-
fidence and is one reason why so many proprietory names
are attached to them. If there is any secrecy surrounding them
it arises only because the industries concerned are reluctant
to disclose their hard-won technologies in face of intensive
competition, often internationally. It should go without say-
ing that the manufacture of by-products follows accepted

and normal standards, often with official inspection, as does basic merchandise.

Technical achievement is one side of the coin, probably the most important, but it is on the sales side that technical accomplishment often fails. This may be illustrated as follows. If the sales demands for some basic material were plotted graphically against a seasonal base it would disclose a recurring pattern of rises and falls. This would suggest that materials becoming available for by-products would fluctuate in sympathy. Likewise, if the demands for the by-product were plotted the times of maximum and minimum would be shown. It is unlikely that the two graphs would undulate in unison. This, of course, is again the vexatious question of supply and demand. A situation could arise when the demand for one is at its highest and the demand for the other at its lowest. This could lead to a state of affairs when there was either a surplus or a shortage of material available for by-products. For some merchandise this would be of little consequence, but in others, especially in the area of perishable goods, the situation would require close attention.

We should not leave this balancing-act situation without again referring to industrial disputes. Without concerning ourselves with the pros and cons of these, it should be clear that a delicate situation would soon arise from stoppages in any industry with strong by-product connections. A example of this occurred in the recent town gas industry dispute. Reductions in gas production immediately resulted in a corresponding reduction in by-products available. So what do the purchasers do? They seek alternatives.

This problem also operates in reverse. A company concerned with the preparation and sale of other people's valuable waste discontinues his orders for this material if his labour is withdrawn. The result is that the waste accumulates at its source and must be stockpiled or destroyed.

At the time of writing the oil crisis is much in evidence and is likely to remain so by the time these words reach publication. It is difficult to say precisely how much industrial production in Britain, and other advanced economies, will be affected by this shortage. But unless there is a sudden improvement in the supply position production must grow less and less. Quite apart from oil for fuel usage, oil products will also fall off quickly. The thing that is quite clear is the reliance of the oil products business on the continuing production of the basic material — oil.

Yet another matter that gives rise to market disturbance is the appearance of new materials, although this is by no means confined to the by-products trade. Many new material have become serious competitors in traditional markets — as their development intended them to be. They are often synthetics derived from new processes.

The by-products industry is largely a matter of separating unwanted materials from the wanted. Mechanical separation is usually not a difficult task and many kinds of machine are used to achieve this end. On the other hand chemical separation is much more specialised. The interested reader is referred to a series of thirteen papers [1.2] covering many of the less common means of separation. The contents of these include some of the latest theoretical and practical thinking by experts in various separation techniques, including thermal diffusion, zone refining, membrane separation, molecular sieves, chromatography and dissociation extraction. Value is added to the high standard of these papers by a full report of the discussion that ensued.

Critics might say that I am overrating by-products by sweeping generalisations, but, of course, achieving effective uses for by-products should not override every other consideration. Clearly manufacturing efficiency rests on the best use of raw materials, power, labour and full use of plant capacity. The pursuit of re-use is imbedded in these, or should be made so. John Philips (1676-1709) summed up the subject accurately enough when he penned these eloquent lines:

> Some, when the Press, by utmost vigour screw'd
> Has drain'd the pulpous Mass, regale their Swine
> With the dry Refuse: thou, more wise, shalt steep
> Thy Husks in Water, and again employ
> The ponderous Engine.

To realise the benefits of and reliance placed on by-products we have only to imagine what would likely follow the sudden abandonment of them.

On the question of nomenclature, one can fully appreciate the difficulties arising when different areas of industry employ the same term to mean different things. Much of the misunderstanding will be resolved with the appearance of a new book, now in course of preparation, covering British pratice on terminology. Its publication is awaited with interest. Meantime, here and there in this book it has been unavoidable to use some terms employed in industry, explanations of a few of which are given below.

Glossary

Anaerobic digestion   Process in which the conditions are provided for anaerobic bacteria to break down organic matter and produce methane. Commonly employed for the treatment of sewage sludge.

Condenser   Apparatus for converting a vapour into a liquid during the process of distillation.

Crucible   A vessel used for high-temperature chemical reactions. A melting-pot.

Decanting   Separation of a solid from a liquid by permitting the former to settle prior to pouring off the latter. Also the separation of two immiscible liquids of different specific gravity.

Distillation   Process for converting a liquid into vapour, condensing the vapour, and collecting the condensed liquid (distillate). The process is performed in a boiler, copper or other apparatus.

Fertiliser   Material used to enrich soils. Products of organic decomposition and waste, manures, etc., containing elements essential to plant life are much used for fertilising.

Effluent   Waste water, usually contaminated, discharged from a treatment plant. Final effluent means a liquid finally discharged after treatment of waste waters.

Electrolyte   A solution that can be decomposed by passing an electric current through it.

Electrolysis   The decomposition of a chemical compound by electricity into its component parts.

Filter   Any porous material or device by which solids suspended in a fluid are removed. Filtrate is the clear liquid after filtration.

Furnance (blast)   Furnace for the smelting of iron from iron ore. It is charged from the top with a mixture of iron ore, limestone and coke. The coke is ignited at the bottom of the furnance by a blast of air. The carbon monoxide so produced reduces the iron oxide to iron. At the same time the heat decomposes the limestone into carbon dioxide and lime. The lime unites with the sand and other foreign matter to form molten slag. The molten iron and slag are tapped off separately at the bottom of the furnace.

Furnace (reverberatory)   A furnace so arranged that the material and fuel to remain separate. Usually the ceiling is heated by flames and the heat produced is reflected downwards to the material below.

Furnance (electric)   Much used for manufacturing calcium carbide (combination of calcium and carbon) and carborundum (combination of carbon with silicon). Also extensively used in steel works, especially for making the special alloy steels containing tungsten or molybdenum.

Greaves (or Graves)   The sediment or fibrous refuse of melted tallow.

Hydro-extractor (centrifuge)   Various classes of machine in which centrifugal force is used to perform separation.

Used for extracting moisture from moisture-laden solids,
cleaning oils containing grit, or for the separation of sandy
substances from broken solids.

Pyrolysis   The term covers the destruction of the natural
form of a material by heating in an enclosed space in the
absence of air.

Re-cycling   The continuous return of unchanged products
to a process, such as the re-use of water. The term is also
used to describe the continuous return of unwanted by-
products to a process.

Refining   The process of separating impurities; to clear of
dross. The purification of a metal from an alloy or other
substance. The raffinate is the refined product.

Residue (residuum)   Materials left after the completion
of a process. Residuum applies to what is left after a
chemical process.

Retort   Any vessel from which distillation takes place. An
autoclave for heating sealed cans by superheated steam under
pressure. A chamber where coal is heated to produce gas;
a furnace in which iron is heated to make steel.

Roaster   Usually refers to ovens and high-temperature kilns
for the removal of moisture from a substance. In simple
form they comprise a heated grate or platform upon which is
laid the material to be dried.

Screening   This is a process for separating solids from solids
on a particle-size basis, or solids from liquids. The apparatus
(many kinds) takes the form of a perforated barrier, station-
ary or moving, through which the material to be screened
passes. Screenings constitute that part of the material left
behind by the screen.

Separation by stratification   Complex mixtures of liquids
may be decanted into individual components by settlement

following storage. The components form into layers according to their own specific gravity value. The layers may be tapped off at the level that they take up in the specific gravity scale.

Solvent extraction   Processes in which solvents are used to dissolve out desirable or undesirable elements from a material The process is widely used to extract oils from seeds and valuable metals from solutions. In oil refining the process is used to remove aromatics from kerosine. In the meat processing trade solvent extraction is adopted for removing grease from flesh.

Straining   This is a form of filtration in which the filtering medium is a membrane such as woven wire-cloth or perforated sheet, whilst micro-straining refers to a proprietary process using a filtering screen of unusual fineness. Strainers may be arranged as drums, discs or bands of straining fabric, the lower part of which is submerged in the liquid so that it can pass through.

Sublimation   The operation of bringing a solid substance into a state of vapour by heat, and condensing it again into a solid by cold.

Tonne   Metric ton: 1000 kilogrammes; 0.9842 ton.

Vitamins   A group of organic substances, occuring in foods, which are necessary for a normal diet.

Winnowing   This is an air-blast process by which certain light substances are separated from heavier ones, such as loose husks from kernels of beans.

References

1.1 *Farm waste disposal: silage effluent,* Ministry of Agriculture, Fisheries and Food.

1.2 *The less common means of separation,* The Institution of Chemical Engineers.

Chapter 2

ANIMAL-BASED BY-PRODUCTS

Primitive man depended for his very existence upon his aptitude as a hunter. The flesh of his quarry provided him with food, the skins his clothing and shelter, and the bones were fashioned into tools; the blood was used in the hunter's religious rites. There was little waste. From these beginnings the animal-based by-products industry emerged.

A few figures quoted from a recent report of the Department of the Environment [2.1] will suffice to indicate the scale upon which the animal by-products industry is carried on today. The report states the industry annually turns out 460,000 metric tons of protein offal meal and tallow, plus 400 million square feet of leather.

The present cycle of events in the recovery of cattle materials is shown in Figure 2.1, although not shown are some supplementary recoveries which I will mention later. A rather similar diagram would also apply to sheep and, on a much reduced scale, to small animals.

2.1 Livestock farm waste

Waste from livestock farms is believed to approach several million tons per annum, most of which is returned to the soil. A few years ago this presented no problem, but recently significant problems have arisen, many of them centred on pollution. Such problems become even more acute on intensive and factory-type farms with little land at their disposal, e.g. complications arise with run-off from milk parlours and winter quarters. In some areas new farm units are not permitted unless adequate waste disposal facilities are provided.

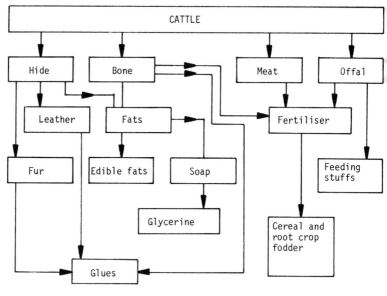

Figure 2.1   Animal-based by-products

Various methods of preparing waste material for disposal exist (composting and lagooning being common) and there is now a trend towards biological treatment, e.g. simple sewage disposal plants, although this may involve solids separation.

Traditional 'muck-spreading' continues, although reducing waste to a slurry and applying it via jets or sprinklers to the ground has advantages. Here again the possibility of pollution arises and it should be noted that direct soakage into the sub-soil may involve a breach of the *Water Resources Act, 1963.* This legislation also applies to the discharge of effluents into rivers, canals, etc. The disposal of farmyard waste into a public sewer has to have the prior consent of the drainage authority.

Much of the above applies to silage effluent which is among the most serious as a pollutant of water. Disposal on land is preferred, and if wilted crop is used there is the advantage of dryer material. The leaflet *Silage effluent* [2.2] furnishes useful information on this subject.

2.2   Meat by-products

This industry is based mostly on the work of the public
abattoir and the private slaughter-house. Ante and post-
mortem inspection decides the eventual destination of any
particular carcase. Dressing involves the separation of edible
parts of the hides, feet, horns, stomachs and other inedibles.
These residuals, as well as condemned carcases, are occasion-
ally processed at the place of slaughter, but more commonly
they are accumulated at private plants where large-scale
operations are carried out more economically. Infectious
and/or diseased carcases are destroyed by burning and do not
enter into the by-products trade.

The trade is controlled by sections of the *Food and Drugs
Act, 1955,* and the *Meat (Sterilisation) Regulations, 1969.*
These regulations, which apply to England and Wales, super-
sede the *Meat (Staining and Sterilisation) Regulations, 1960,*
as amended, and came into force in 1969. The second part of
these regulations applies to unfit butchers' and knackers' meat.
The third section deals with imported meat that is unfit, or
not intended, for human consumption. The fifth part covers
administration and general matters.

The processing of abattoir residues is largely directed to
fat recovery and the manufacture of feeding stuffs and
fertiliser. Some methods are given below and these are followed by reference to lesser retrievements. The word 'rendering'
herein signifies boiling and clarifying. Leather is dealt with
separately.

2.2.1   Wet and dry rendering

The simple process of wet rendering entails cooking the
material in a vented vessel heated by a steam jacket. Fat is
separated, although a considerable amount of protein is lost
with the waste liquid when the vessel is drained. The residue
solid matter goes as fertiliser.

In dry rendering the main function of the vessel is to reduce the moisture content of the material and to liberate fat
which acts as a heat-transfer medium so that the charge under

treatment is thoroughly heated. By using this method a higher
fat yield is obtained than in the previous method. The residue
is suitable for a feeding meal.

2.2.2. Pressure rendering

In this method steam pressure is applied to the vessel and its
jacket instead of directly to the material. This results in an
early breakdown of the material, thus eliminating the need
for size reduction beforehand. The collapse of the fat cells
under steam pressure gives a higher grade of fat than that
resulting from non-pressure rendering, whilst the resulting
meal is unchanged.

2.2.3   Subsequent processing

On completion of the cooking or rendering process the charge
is delivered into a tank with a filter bottom. Through this a
large amount of fat is drained out. The greaves are then pressed or centrifuged to remove more fat. The two lots of fat,
each containing water and sludge, are brought together and
allowed to settle and separate. The fat thus separated is then
ready for marketing. The greaves are ground into a meal,
screened and disposed of as meat-and-bone meal, bone meal,
or fertiliser, depending upon the type and quality of the raw
material.

2.2.4   Solvent extraction

In this process the material is dehydrated by indirect steam
heating in a vessel. Fat is extracted by introducing a pre-heat-
ed solvent which unites with the fat in the material to form a
concentrated miscella which is removed by pumping. Miscella
removed from the vessel is settled in tanks from which the
highly concentrated fraction is passed to an evaporator. The
weak miscella from succeeding washes is re-circulated to form
initial washes in subsequent operations. The residual solvent

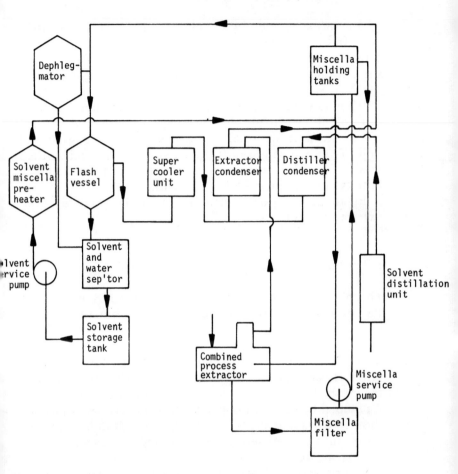

Figure 2.2   Process flow diagram of combined solvent extraction
plant. *Courtesy:* Iwel Engineering Limited

in the degreased material is removed by steam stripping, and
the final discharge is a dry friable product suitable for cooling
and grinding.

The fat content in the concentrated miscella is recovered by
distillation of the solvent in the evaporator mentioned by
indirect steam heating. The solvent passes through a condenser and is then suitable for re-use. The fat is discharged to a
clarifying tank.

An alternative method of solvent extraction is by centrifuge. In this the greaves are packed into the machine's perforated basket and spun. Hot solvent is pumped up through the central hollow spindle and disperses centrifugally through the greaves.

When raw materials are fresh and are of the poorer portions of edible carcases the rendered fat is classified as edible. When the materials are of an inedible nature, or contain stale or decomposed edible stuff the recovered fatty-acid content may be high. This material is termed 'technical fat' and it finds outlets in the soap and lubricants industries.

Figure 2.2 is a flow diagram of a typical combined solvent extraction plant.

2.2.5   By-products installation at Sheffield

This plant employs the combined cooking and solvent extraction process for the production of low fat content feeding meals and technical fats from a variety of abattoir residues.

Raw materials are trucked to the plant where, after classification, they are fed into chutes, bones and carcase material into a crusher and washed offal to a pump. Conveyors elevate the crushed items into a storage hopper at roof-level. The offal is pumped to another hopper at the same level, first passing through a draining drum over the hopper. Both hoppers discharge into two rendering vessels equipped with steam heated jackets and agitators.

The raw material, after a period of pre-heating, is dehydrated to release free fat and then solvent extracted by a number of washings, beginning with weak miscella. This reduces the fat content of the solid material to between 3 and 5 per cent.

After extraction, the residual solvent is removed in vapour form by indirect stripping and the solids discharged as a free-flowing mass, via a magnetic separator to remove tramp iron, to a pulveriser and thence to a grinder.

A feature of the plant is the removal by fan of hair and the sifting of over-sized meal particles to obtain a constant grade of product. The meal is then passed into a multiple spout conveyor for bagging ready for sale.

Fat *arisings* removed as a miscella are treated for the removal of solvent by the same means as already described. The fat is then gravitated to a settling tank for final clarification.

Cooking odours, always a problem, are reduced by vapour handling in surface condensers. These remove the latent heat of the effluent, which after being condensed, is passed through a final cooler and thence to a separator where the solvent is isolated from the condensed effluent, the solvent being returned to a storage tank and the effluent to a drain.

2.3 Other by-products

Hoofs and horns may be rendered with the main body of the and separated manually afterwards. But because of their toughness they prolong treatment time. If available in quantity they are treated separately. Grinding into powder results in a slow-acting fertiliser.

The use of ground bone as a fertiliser commenced when eighteenth-century workmen observed that crops grew well on land where bones had been stored prior to their manufacture into cutlery handles. Calcined ground bone is used in the pottery industry to make the finest bone china. Unclassified bone is ground to form a fertiliser.

Basic materials for glue are bone and the inner structure of toe and horn. High class glue is made almost exclusively from mature bone, whilst an indiscriminate mixture gives an average product.

In the manufacture of bone glue the material is crushed and subjected to treatment by a vaporised solvent in a vessel. This drives off moisture and dissolves out fat. This is followed by applying superheated steam to remove the solvent and moisture. The degreased bone is then removed from the vessel in a comparative dry state. The fat drainings from the vessel and the mixture of vaporised water and solvent are then cooled and separated. The fat is mostly used in the manufacture of soap, although glycerine and fatty acids are also by-products. The degreased bone is refined and subjected to treatment in an autoclave to produce the end product.

Detailed information on glues is given in *The Story of*

Animal Glue [2.3] .

Catgut for musical instruments is formed from animal intestines twisted into strings. Its latest application is for high torque flexible shafts for power transmission.

The traditional outlet for entrails is the sausage casing trade, although synthetic casings are now a rival. The natural casings are derived from selected material, cleansed and sterilised.

Various classes of gland are collected from large slaughter-houses for use in many pharmaceutical products.

An interesting paper [2.4] submitted by Rubin to the Agricultural Institute of Alberta gives a picture of the present utilisation of glandular materials for the preparation of fine chemicals and biologicals in Canada. It is pointed out that the meat packinghouse industry in that country is unique in that it has available in one location a vast quantity of glandular materials so that arrangements can be made to process them economically. This is not the case in Europe where, by and large, packinghouses are small and widely dispersed. It is for this reason that the Canadians can consider the production of fine chemicals and pharmaceuticals from packinghouse by-products.

In Canada the return from by-products has fluctuated a great deal and has not always kept pace with rising costs. Thus, the return from hides varies widely due to the fierce competition from synthetics that leather has had to face. There has also been a very considerable decline in the glandular field. Of all of the glands available from the animal carcase about eight are collected regularly in Canada. Advance in science have supplied the process people with better and more certain sources of active principles in the case of many of these glands. Quoting Rubin:

> 'We built a Fine Chemicals Plant a few years ago in Toronto in which our operations are centralised. Thus, we hope to be able to reverse the trend and to get a bigger return from animal carcases via fine chemicals. . . We have since moved on to other packinghouse raw materials in search of products of pharmaceutical and industrial interest. We make pepsin from hog stomach

linings. Pepsin is a digestive enzyme which is used as a pharmaceutical product. From the stomach of young calves we isolate rennin, the milk clotting enzyme. This enzyme is very important to the cheese industry. . . .

'Bile has been of great interest to us for a good many years. From beef bile we make cholic acid which, by a simple chemical conversion, can be transformed into dehydrocholic acid, a popular digestive acid. From the same source we get desoxycholic acid which is an important raw material for the production of cortical hormones. Other raw materials, such as those derived from Mexican plants, are alternative materials, but desoxycholic acid from bile has never been pushed out. . . .'

The paper also points out that from hog bile other interesting bile acids can be made whose potential is still to be realised.

It is widely known that the drug insulin is made from the pancreas gland, and in Canada this job is left to Cannaught Laboratories of the University of Toronto, the place where the drug was discovered. The pancreas is, however, a source of enzymes and, apart from insulin, trypsin and chymotrypsin are possibly the most important. Quoting again:

'There are other materials of glandular origin in which we are interested, such as desiccated thyroid and heparin . . . . Heparin is nature's own anti-coagulant. It is made from lungs and casing slime. It is often used in the treatment of heart disease. . . .'

The following glandular materials are still being offered commercially — hog and beef pancreas, hog and beef thyroid, beef bile, superenals, hog stomach linings, calf vels and testes.

Slaughterhouse blood is allowed to accumulate in a catch pit, from whence it is either blown by steam or air to a container for eventual transportation to specialised processing plant. Here the fluid, which contains much moisture, is coagulated by live steam injection, followed by drying to form a powder. Alternatively, the fluid may be placed direct

in a dryer, but the process takes longer. Although the powder
has a high protein content, it is usually disposed of as a
fertiliser.

The Meat Research Institute is currently devoting considerable attention to blood collection and utilisation, including
its use for human consumption. Much is expected from these
researches.

2.4   Tanning industry

It is only necessary to give a résumé of the processes involved
in this trade, for they are fully traced in general works.

The chief stages are given below, although all may not be
applied to a particular type of skin or hide.

2.4.1   Before tannage

Removal of the skin from the carcase. Preliminary curing to
preserve the skin. Washing to return the material to its former
state. Liming to free fat, flesh and hair prior to tanning.
Removal of these materials. Deliming to neutralise the alkali
from liming. Bating to soften the skin. Pickling to correct
acidity for tanning.

2.4.2   Tannage

The tanning of the hides and skins by appropriate methods,
employing vegetable tans, chrome, alum or oil, etc.

2.4.3   After tannage

Heavy Leather   Washing to remove surplus tan. Flattening.
Oiling to acquire flexibility. Waterproofing by impregnating
with oil and fat. Drying. Boiling.
Light Leather   Splitting to achieve uniform thickness. Washing to remove surplus chrome salts. Neutralising to correct

acidity. Dying to give the required colour. Fat liquoring to give softness. Removal of creases and surplus moisture. Drying. Staking for flexibility and softness. Glazing, plating and embossing to customers requirements.

The results of these steps in tanning gives rise to a series of useful by-products, but further treatment usually ensues. With the exception of effluent, which is again referred to, useless waste is minimal. Hair has a ready outlet to firms manufacturing felt or cheap blankets. It is also used as a stuffing for upholstery and as a binding agent in plaster board for the building trade. Pig and hog bristles are disposed of to makers of certain classes of brush. Limed hide trimmings and detanned chrome leather shavings are retained for making gelatine. Sundry waste, e.g. buffing dust and leather scrap, may be used as a fertiliser. Leather pieces and trimmings are absorbed by the manufacturers of leather buttons, toys, purses and the like.Remnants of this sort are also supplied for the patching and repair of footware. Natural fat arisings are used for soap production, and horns and bones are traded to merchants for eventual grinding into bone meal. Chamois leather has good filtering qualities and off-cuts and odd pieces are sought by makers of this kind of equipment.

Effluent from tanning operations is complex and there are a number of methods of handing this. Many tanneries treat their own effluent to an acceptable degree, which means that the effluent enters the public sewage system at a quality no worse than that of household sewage. The only problem is one of odour and this is slowly being solved. The final sludge from the effluent pits makes an organic fertiliser and this has also been used to prepare vitamin B12.

2.5 Fur

The fur trade in contrast with the tanner has the prime object of keeping the hair securely on the skins it uses. The process involves a preparation that removes excess grease and adhering dirt from the hair followed by a modified form of tannage which preserves the leather without damaging the appearance or properties of the hair.

Only skins with relatively fine silky hair are suitable for furs and thus sheep and lambskins are materials mostly used. During the period 1951-54, however, prior to the incidence of myxamatosis in the rabbit population many thousands of wild rabbit skins were imported from Ireland under refrigeration by the fur trade from the Dublin rabbit meat packing plants. Two examples of the uses of skins are:

1   Sheep, lamb and beaver skins, long wooled lamb trimmings, slipper trimmings, bootee and shoe sheepskin linings, sueded sheepskins for coats, hats, gloves and slippers, polishing pads, nursing sheepskins, rugskins and powder puffs.

2   Rabbits for furs and trimmings.

Most of the other skins used by the trade are either bred on farms or trapped for the specific purpose of fur production and are not in any sense by-products of other industries, although mink farms use meat offal and fish filleting wastes to feed their stock.

During the course of processing a number of waste materials arise that can be of recovered as by-products. Raw skin trimmings are usually too small to be processed and are often sold to glue and gelatine manufacturers. Unwanted membranes on the back of the skins which are scraped away to allow dressing liquors to penetrate through the skins, are mostly rich in grease which can be extracted.

Surplus wool and hair is also usually saleable and is collected, washed, dried and baled.

These by-products can only be an economic proposition if sufficient quantities of satisfactory quality can be accumulated quickly enough and the cost of transport can be recovered. Conditions favour the larger factories and those in traditional production centres, e.g. Northampton.

Effluent from the processing of fur is generally in line with that emitted by tanneries and its treatment before discharge is the same.

2.6 Wool

As will be seen from Figure 2.3 most of the by-products arising during the various processes of the wool textile industry are fed back into the woollen section of the industry at the carding stage.

The raw wool brought into the mills for processing, is, after sorting, scoured to remove all dirt and natural wool grease prior to the wool-combing process which lead eventually to the production of yarn. The traditional way of scouring is by continuous agitation in a solution of hot water to which has been added soap and soda. The resulting dirty water is then removed by wringers. Some of the liquid, the less contaminated, is returned to the system for further use. To this is added fresh solution to make up for the loss. Usually the spent and dirty liquors are run to settling tanks to allow sandy material to settle. This solid material is often dumped. The liquors or effluents are treated in one or more of the following ways:

1 Direct discharge to the public sewer system (as at Bradford).

2 Centrifugal extraction of the globules of lanolin for use in the cosmetic and pharmaceutical industries, the liquid being discharged to the sewers.

3 Extraction of wool grease by treating the liquors with sulphuric acid. This destroys the soap (previously inserted to assist scouring) and precipitates the grease and other emulsified solids which are deposited as sludge. This is removed, boiled and filter-pressed to yield the wool grease. The semi-clarified liquors are then discharged to the sewers for treatment as in (1). The grease obtained is used for lubrication and rust-prevention purposes.

The Bradford Corporation accepts into the sewers the waste from the industries of the city, and a substantial proportion of this originates with the local wool scouring industry. The works at Esholt were specially designed to deal with

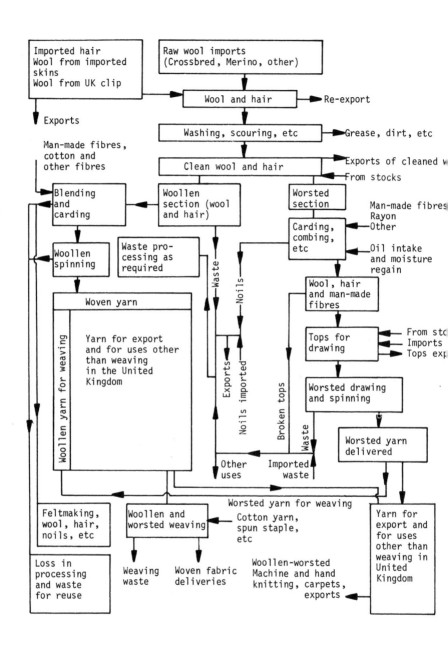

Figure 2.3 Production flow chart for the wool textile industry.
Courtesy: Wool Industry Bureau of Statistics

this strong waste, the basis of the treatment being acid cracking of the grease emulsion using sulphuric acid, the acid being manufactured on site.

The grease in admixture with the sewage solids is passed to the sludge treatment plant where it is conditioned, by adding a further quantity of acid, and boiled with live steam. The sludge is then put into filter presses where the grease is melted out by steam. The filtrate and grease are passed to the by-products section for processing into saleable collaterals.

Wool grease is used in the preparation of temporary corrosion preventives, waterproofings, special lubricants, putty, mastics, certain soaps, marking fluids and leather preservatives, although these items do not exhaust the long list.

Lanaloid obtained from the process is used as a compounding agent for increasing body and oiliness or for introducing gelling and thixotropic characteristics into oils. It is also employed to form tenacious plastic films suitable for application to industrial or other heavy fabrics, and which have good water proofing properties. Paraffin wax may be included in mixtures to enhance water repellency. Another use is as a pigment binder for crayons and carbon papers. But the economics of the material becomes more difficult as markets shrink.

At Bradford, as at various other places, the pressed cake resulting from grease removal, is tipped on land and left to mature. Grinding then follows to form an organic fertiliser.

Gardner [2.5] discusses the Bradford plant in detail, and a publication by Truter [2.6] deals with wool wax, its occurance, function, chemical composition, recovery and fractionation technology.

2.7 Poultry industry

Intensive poultry rearing and battery type egg production have brought a set of problems hitherto of little consequence. Some of these − waste disposal being one − are overcome by methods applied in other departments of modern farming.

Certainly there is considerable detailed information on this matter that has appeared in the technical and commercial press.

2.7.1   Liquid manures

Liquid and semi-liquid matter arising from intensive egg production contains anything up to 85 per cent moisture and it is usually disposed of by spreading over land or dumped at disposal points. Both operations need to comply with the requirements of local authority anti-pollution measures.

2.7.2   Deep litter and built-up litter

These materials arise from breeding and broiler operations, as well as intensive egg production where birds are kept on sawdust, wood shavings, etc., which ferments and composts much of the manure. The moisture content has been estimated to be anything between 15 and 30 per cent of the whole. It is used as a fertiliser, but a better outlet seems likely when the material is dried. This was anticipated in the USA as long ago as 1954. Oliphant's paper [2.6] refers to more recent experiments in this field and makes reference to feeding trials in this country. It appears likely that the process will develop as the demand for protein increases and the technical merit of nitrogen recycling becomes accepted.

2.7.3   Processing by-products

The by-products of poultry processing, estimated to amount to between 20 and 25 per cent of the intake waste, consists of feathers, blood and inedible offal. The materials are either disposed of to local animal feed outlets such as mink farms or may be processed either separately or in combination to give high-grade protein concentrates suitable for use as animal feed ingredients. Further, on the large-scale plants oil extraction is carried out since the de-oiled meal has a better keeping value and commands a better price, and the purified fats find a ready market.

The more minor sources of waste such as hatchery clears, incubator debris, etc., are often disposed of in the simplest way possible in view of the relatively small quantities involved.

Apart from the many substitutes now available for feathers and down, there is a sizable industry dealing with these by-products of poultry handling. They are collected, sorted, washed, dried and sterilised. The finished materials go to manufacturers of cushions and bedding for stuffing, as well as to the fancy goods trades where whole feathers are used. There still remains an outlet for whole feathers for making quill pens, shuttlecocks and floats for fishing tackle, although the demand is diminishing.

For some years feathers have been converted into animal feeding stuffs but there is only a small amount of feather residue (unsuitable for feeds) that reaches the fertiliser trade where the nitrogen content only is of value. Because of more suitable materials the value of feathers for this purpose is limited.

A rather surprising departure is the employment of feet of ducks for human diet. One of Britain's largest firms of duck breeders is exporting to Hongkong six million duck's feet annually for human consumption. This kind of unusual trading suggests markets yet to be exploited. The growing popularity of Chinese cooking might well extend this old custom to Europe. It further implies it would be worth taking a good look at other foreign cooking practices with a view to absorbing other edible materials now barred by custom or for religious reasons.

Eggs are graded into sizes (and sometimes into colour as well) most attractive to the housewife and restauranteur for direct table use. By and large the remainder are absorbed by the few companies concerned with proprietory brands of prepared foods.

Very approximately, 80 million dozen eggs are processed each year by the egg products industry, and in recent years efforts have been made to find a use for the 5000 tons of shells that are produced as a by-product. Investigators have shown that true shell is almost completely calcium carbonate. However, the shell membranes and adhering ablbumen give commercial shell a protein content of 6 to 8 per cent (75 per cent digestible). Adhering albumen can contribute as much as 20 per cent of the wet "egg shell" weight. It seems that egg shells are not likely to be widely used in animal feeds because the available amount is small, the protein value too low and the cost of limestone as a source of calcium carbonate is comparatively cheap. The Eggs Authority have pioneered work on the use of shells as an abrasive in compounds used for cleaning brick, stone or metal work.

2.8   Dairy Industry

Most of the milk produced in this country is required for liquid consumption and it is mainly during the period of peak production in the spring that supplies are available for the manufacture of dairy products.

The chief products are butter, cream and cheese, and these are the spearhead of subdivision materials indicated in Figure 2.4.

To consider all in detail would demand considerable space: and indeed this has already been done more than adequately by Davis [2.8]. It therefore remains to highlight a few main points.

The amount of home-produced butter is estimated to be never more than about one-tenth of the total consumed; the remainder is imported. Dairy products are manufactured in the chief producing areas, notably in the West Country, the milk being brought to depôts where processing takes place. Its composition varies according to the type of milk and the seasons of the year.

In the manufacture of dried milk the natural water present is removed, the method selected differing with the individual manufacturer concerned. British regulations require dried whole milk to contain a minimum of 26 per cent butterfat.

Cheese is made by coagulating milk with the aid of rennet. The curd so formed is drained and sometimes pressed and

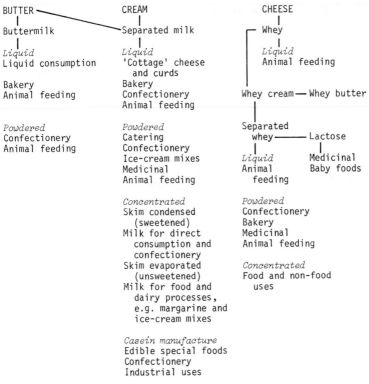

Figure 2.4 By-products of butter, cream and cheese. *Courtesy:* Dr J.G. Davis

allowed to mature. These basic principles are the same for all varieties.

Whey, which occurs during cheese-making, contains most of the milk sugar or lactose and the water-soluble protein of the original milk. The disposal of this material was a problem and much of it was fed to pigs. Now it is condensed and the lactose extracted for some industrial and pharmaceutical purposes.

As in the case with many other trades, effluents arising from dairy activities may not be discharged direct into water-courses, sewers, or, in some cases seeped into the ground, without positive planning and prior approval of the authority concerned. A Government publication [2.9] gives advice on this.

2.9   Fish processing industry

The by-products originating from fish are mostly concentrated on the production of oil, animal feeding stuffs and fertiliser. Since these much-demanded materials are mostly imported there is every reason for making the best use of home sources. The programmes of the fish trade research organisations confirm that retrieval is very much in mind. The trade looks to the future by way of The Association of Fish Meal Manufacturers' nutritionists working in collaboration with the Department of Scientific and Industrial Research. This kind of cooperation lead to the discovery that the traditional pemmican used to feed sled dogs on Polar exploration was, so the claim goes, not as suitable as a new diet based on white-fish meal. This diet was used on the Trans-Antarctic Expedition led by Sir Vivian Fuchs. The latest research, undertaken by United Nations' specialists, suggests that fish 'flours' can be used as sources of protein in human diets. We have yet to see the full development of this discovery.

2.9.1   Fish meal

The following is an extract from a bulletin produced by the Ministry of Agriculture, Fisheries and Food [2.10]:

> The superiority of the animal-protein foods, such as fish meal, meat meal, etc., over the vegetable-protein foods, such as extracted decorticated ground-nut meal and extracted soya bean meal, for the nutrition of pigs and poultry has hitherto been attributed solely to the better balance of animal protein in respect of the amino-acids needed for building up the body protein. Recent research however, has shown that this is not the whole story, and the conclusion has been drawn that animal-protein foods contain small amounts of a growth factor to which the name 'animal-protein factor' has been given.

The presence of this growth factor is held responsible for the

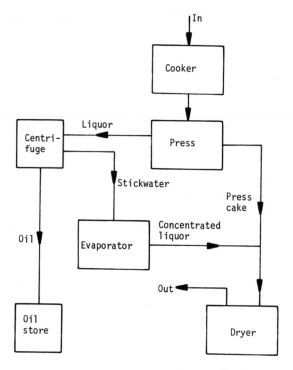

Figure 2.5 Flow diagram of fish meal plant

superior growth and reproduction in animals maintained on
rations containing crude products.

Almost any fish from the sea can be used to make fish
meal but we are concerned with waste. In the UK it is
manufactured from unsold fish, heads, skeletons and trimmings. There are several methods of manufacture, the following being quite usual. In this the raw items are cooked,
dried and ground in that sequence, the plant being arranged
as shown diagramactically in Figure 2.5. Cooking is carried
to a stage at which the material is suitable for pressing.
During this second stage most of the oil and much of the
water is squeezed out, leaving behind solid matter — the press
cake. The expressed liquor passes to a centrifuge where the
oil content is separated from the watery liquid — the stick-

water. The oil is conveyed to storage receptacles for eventual refining.

In this country oils are widely used in the manufacture of edible oils and fats, for example margarine. USA regulations do not permit fish oils for human consumption, but are much used in the varnish and paint trades.

The stickwater contains dissolved material and small solids. The latter are mostly protein and the water can hold considerable fish particles. Because of this the effluent is subjected to an evaporation process which results in a syrup known as condensed fish solubles, and these are marketed separately. Sometimes this concentrated liquid is fed back into the press where it rejoins the press cake. The press cake is finally dried and ground to form the meal already referred to.

Condensed fish solubles represents a useful source of protein to the stockkeeper particularly when supplies of protein materials are limited.

Another by-product of the fish processing trade is photo-engraving glue used by photographers. It is made up of fish glue, water and ammonium bichromate.

It would be as well to mention that fish, excluding shell-fish, fall under two headings – white fish (non-oily) and red (oily). In the former group are whiting, sole, cod, etc., and in the latter salmon, mackerel, herring and a few others.

Windsor discusses fish meal [2.11] and fish protein concentrate [2.12], whilst an explanatory leaflet [2.13] deals with aspects of condensed fish solubles.

2.9.2   Cod liver and cod roe

Cod gutting and liver separation operations are usually completed at sea during the homeward journey of the trawlers. This is irrespective of the distances involved. In the case of long passages, however, rendering equipment is installed aboard the vessel and by this the liver oil is liberated from the tissue and delivered to holding tanks. From these it is decanted into drums prior to landing and final handling on shore. The individual drums are sampled and the material allowed to settle. When short sailings are involved, rendering operations are completed at shore-based plant.

If the oil is intended for medical purposes it is required to fulfil stringent conditions, one of which requires filter pressing in a cold room to extract the stearin. The material is then deodorised and held in a finishing medical receptacle against orders. A three-hour duration standing clarity test at 0°C is also imposed.

If the oil is destined for animal feed outlets the cold filtering operation is omitted unless specifically called for, as in the case of winterised veterinary cod liver oil.

Completely processed cod liver oil is almost 100 per cent fat, and possesses high nutritive qualities. Among its uses is that of a canning oil. It can be used in cookery, through owing to its slightly fishy flavour is more suitable for fish dishes.

Cod roe is separated out, selected for quality and marketed raw or as a cooked product.

2.9.3   Isinglass

This is derived from the air or swim bladder of certain fish, namely the teleosti or vertebrate group. The bladder is mostly found attached to the backbone of the fish and on removal it is cut open, washed in fresh water and left to dry. Isinglass manufacturers then process the material in various ways. This may involve soaking and feeding into rotary cutters to produce strips of varying thickness. Isinglass is insoluble in cold water; it simply swells and takes up some water. Upon heating it dissolves, and on cooling forms a tough substance resembling hide or bone gelatin. Most isinglass is not heated now but soaked in an acidic solution, usually tartaric acid, until it dissolves. The final preparation is used for clarifying beer, wine and certain liquors, spirits and other fermented beverages.

2.9.4   Fish shells

One does not foresee a great extension of this minor branch of the fish processing business although there seems to be latent possibilities. Crab, crayfish and lobster parts, where available in sufficient quantity, can be rapidly dried and ground to a grit which is valuable for poultry feeding, and contains a percentage of protein. Cockle and oyster shell can also be treated, but are mostly composed of lime.

There is another side, a waste recovery operation, to shell-grit production. Shoals of shells, usually of the same group, are regularly deposited on certain areas of the littoral. Some of these deposits have existed for ages and they have the advantage of seasonal replenishment by the action of tides. Longshoremen still grind these shells on site for the 'chicken-grit' trade.

The hard, silvery, iridescent internal layer of pearl-shells is removed and marketed as Mother-of-pearl. There is still a lively business carried on in utilising this material, mostly for ornamentation work.

Pearls are formed within the covering of shellfish, especially oysters. When found incidently they are the most valuaable of all by-products. Purposely cultivated pearls are the end product of a separate industry and are not by-products.

2.9.5   Smoked salmon industry

Salmon have been smoked in various ways for hundreds of years. The Department of Scientific and Industrial Research publish a leaflet [2.14] that gives the general principles of smoke-curing of fish and from this much useful information may be gleaned.

In this industry there is very little of the fish that is not used in one way or another. Offal — mostly heads, gut, bone and tails — is usually processed to form an animal feeding stuff. Otherwise there appears to be little interest shown for the further use of this residue.

2.9.6   Fish canning

There is considerable waste resulting from canning of fish and, in general, this does not appear to be utilised to a great extent.

This may be due to the fact that in many cases canning factories are in inaccessible places such as Alaska. It appears that unwanted cannery debris, a likely source of raw materials, is much under-exploited and awaiting development. But recycling is not always an easy answer and there are economic factors to consider.

2.9.7 Sponge

The use of sponges has gradually receded in favour of synthetic varieties, mainly because the latter are cheaper and it is unlikely that natural sponges will be able to retrieve their former markets. Only in the leather and pottery industries does the natural material hold its own. The sole by-product originating is during shaping, the off-cuts being used as cosmetic sponges and these have a preference over other materials.

2.9.8 Seals

The annual culling of seals in this country is primarily aimed at keeping the seal population at a reasonable level. In many other areas, however, wholesale killing is a purely commercial proposition. These mammals, especially the pups, are prized for their soft skins. When slaughtered their bodies are suitable for animal feeding stuffs and fertiliser, altbough they may not always reach these destinations. Perfect skins are worth about £40 each to the hunter and are, after tanning, used mainly for fur coats and handbags.

This is an industry with little control over its activities and could well be in sight of its demise. Unlike planned domestic cattle breeding, there seems to be little serious regard to conserving the species.

2.10 Sewage

One of the statutory duties of a local authority is to receive and dispose of domestic sewage in an efficient manner. Since most also accept a large amount of effluent and washings from private enterprise it follows that the total matter piped to public sewage disposal works is not entirely of animal origin. This point, however, would be a fitting place at which to mention how the sewage service has penetrated the business of by-products. The reader is referred to the mass of published information on disposal treatment. It is sufficient to say here that all methods produce effluent, sludge and sometimes methane gas. A brief understanding of what is involved can be gained from the simplified flow diagram shown in Figure 2.6, which, in this case, refers to the Aylesford works belonging to the Borough of Maidstone.

2.10.1   Sludge

The terminology of sewage sludges is explained by Lewin as follows:

> Raw sewage sludge means the sludge from primary sedimentation of raw (crude) sewage and is probably meant to indicate that secondary sludges produced during biological filtration have not been returned to the inlet of the works for settlement along with the primary solids. Thus, if secondary sludge is recycled it then ought to be referred to as mixed primary sludge and if secondary sludge is not recycled then it is as well to define it as raw sewage primary sludge. The secondary sludges, that is the waste bacterial cells called surplus activated sludge, or if from percolating filters, humus sludge, is of course of a more uniform character and is frequently dealt with separately and because of the fine particle size normally requires three to four times more chemical coagulent to enable them to be de-watered. Digested sludge is a term used in referring to sludges that have been anaerobically digested or fermented, usually at elevated temperature for up to 30 days and hence 50 per cent of the original organic matter is degraded to volatile fatty acids and eventually

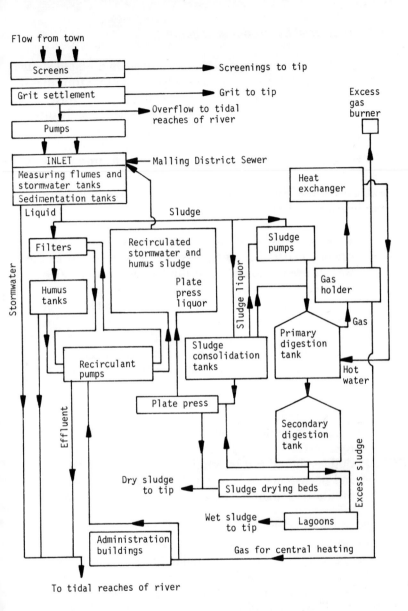

Figure 2.6 Simplified flow diagram for sewage treatment.
Courtesy: Borough of Maidstone

methane and carbon dioxide. Here again digested sludge may mean digested mixed sludge or as at Oxford digested primary sludge. Sludges are often kept separate because the cost of chemical coagulent demands this economy.

Concerning the use of sewage sludge as a fertiliser and its value as manure, some comments are given in the following extract from a Ministry of Agriculture, Fisheries and Food publication [2.15]:

Raw, or digested, sludge is available as a source of organic matter in some areas. At the sewage works the wet sludge is allowed to dry in lagoons or drying beds exposed to air, which produces a raw sludge with a moisture content of 30-50 per cent. Digested sludge is produced from raw sludge by a process which reduces its bulk and results in a product that is easier to dry.

Sludge is a source of organic matter giving some nitrogen and a little phosphate. The organic matter is different in its nature from the strawy fibrous material in farmyard manures and there may be a high content of heavy metals, e.g. chromium, nickel, zinc and copper, in sludges from industrial areas, and these induce toxic effects on crops. Large annual dressings of these sludges should not be applied because of the buildup and persistence of heavy metals, and technical advice should be sought where there is any doubt.

In addition to the forms of sewage sludge mentioned in this quotation, in recent years some local authorities have been applying liquid digested sludge to farmland by means of tankers, this material having a high water content.

A notable exception to using sewage sludge is in lettuce growing. The reason why direct application should not be applied to this plant is the possibility of direct contamination of the leaves (which are of course later eaten) both by disease organisms such as *Salmonella* or by toxic materials in the sludge. There are no risks of this kind where suitable sewage sludge is applied to the bare soil before sowing lettuce but only when the sludge, especially if used in liquid form, is applied over the crop.

On the question of toxic elements mentioned in the above extract, an Advisory Paper by Chumbley [2.16], now being revised, gives recommendations for maximum safe levels of application of certain heavy metals. The Water Pollution Research Laboratory does issue in their series of notes on water pollution a leaflet [2.17] summarising the matter of sewage sludge in agricultural applications quite clearly. Further pertinenent information is contained in the 'Jeger' report [2.18] and a leaflet [2.19] on water pollution.

It is estimated that about 60 per cent of sewage sludge is not utilised on agricultural land. About half of this is dumped at sea, most as liquid digested sludge, and the remainder is dumped on land, generally after dewatering and in a few cases also after incineration. A small amount of liquid sludge is incinerated in admixture with domestic refuse. To the uninformed there seems more than a little room for action in this particular field, but the problem is frought with difficulties.

2.10.2 Sludge gas

It will be seen in Figure 2.6 that sludge is delivered to a special tank and during its retention there certain bacteria act on the waste, multiplying and discharging methane as a product. This is an odourless gas used as a fuel. The process is accelerated by heating the sludge. The heat may be supplied by boiler plant and circulated through hot-water pipes within the tank. The aim is that once the process commences the plant become self-supporting by using the methane as a fuel. Either a single boiler, dual-fired, or two boilers, one to burn oil and the other to burn gas, may be used. In the case of the former the boiler is commissioned on oil and the sludge digestion commences, the gas being tapped off and fed to a holder. As the holder fills with gas and rises it trips a switch which automatically closes off the oilfiring and switches to gasfiring. The plant then continues to operate on sludge gas until there is insufficient gas available, whereupon a further switch operates to change combustion from gas to oil. In the case of the latter a similar set of operations is followed, whereby oil and gas firing alternate as circumstances dictate.

The ability of the dual-fuel internal combustion engines to operate with minimum liquid pilot fuel injection can often be utilised to advantage. This is shown to be the case in sewage disposal works where plentiful supplies of sludge gas of high methane content are to hand. The fact that waste heat from the cooling jackets of these engines is available for use in the sludge treatment process means that a high degree of fuel utilisation is possible. Another reason why the dual-fuel engine is suited to such work is that in the event of failure in the supply of sludge gas, the engine can revert to oil operation.

Usually these dual-fuel engines are coupled to electrical alternators to furnish power for operating the sewage disposal plant and other duties. To give an example reference is made to a large sewage disposal works on the outskirts of London. The normal daily power requirements here amount to about 1170 kW, while maximum conditions need 2000 kW. These demands are within the capability of engine-alternator sets operating on sludge gas generated on site. The engines are cooled on a closed circuit basis, the pumps delivering the outlet water through exhaust gas boilers to heat exchangers. A secondary circuit connects the heat exchangers to the digestion tanks, thus furnishing the process plant with a hot water supply.

At the Katherines Farm Sewage Works, Bristol, there are three nine-cylinder dual-fuel 1240-kW engine-driven generator sets with four eight-cylinder 494-kW sets. Fuelled by sludge gas these engines make this large complex entirely self-supporting from an electrical power point of view. The sludge gas engines at Bristol are shown in the Frontispiece.

2.10.3   Grit, sand and flotsam

These by-products are worth mentioning although they have

ittle value. They are screened out of the crude sewage on
arrival at the works. Sand and grit, when washed, is suitable
as a land-fill and the floatable debris, when mixed with
lime, may be spread on the land. Two more recent developments in dealing with flotsam are gaining favour: one is to
macerate the flotsam and return it to the system and the
other to press out the liquid and feed the semi-dried debris
to a sludge-gas-fired incinerator. This second method results
in a small amount of inert gas to be disposed of.

2.10.4   Effluent

Sewage effluent must be accepted into the environment some-
where and dispersed in such a way as to produce minimal
pollution or ecological disturbance. Its usual destination is
natural water although before discharge takes place the
effluent must be of a standard required by various
legislative enactments.

Much as been said of the increasing cost of water and the
mounting restrictions on its use. These considerations raise
the question of recycling in industries where potable water
is not of great importance. Examples arise with fire-fighting,
cooling and hydraulic pressure systems.

An approach to recycling sewage effluent is being made at
a municipal sewage works in the north of England. At this
site a further treatment plant has been installed from which
waste water is pumped to a nearby woollen mill for re-use
in textile wet processing. This is part of a project being
carried out on behalf of the Department of the Environment
in order to examine the feasibility of re-using in mills the
waste water normally discharged by sewage works. A paper
by Hirst and Rock [2.20] describes in detail this project,
and this is enhanced by a discussion.

Whilst it is safe to assume further advances in this field
are likely, it is probable that uses other than re-cycling will
be evolved for sewage effluents. One forward-looking proposal suggests its use as a culture, and the Woods Hole Ocean-
ographic Institution in the USA is actively engaged in this
field. Behind this project is the employment of tertiary,

or triple treated sewage as an environment in which to cultivate oysters. Oysters are not a common feature of diet, but if sewage-aided farming becomes acceptable the position is likely to be reversed.

2.11   References

2.1   Report of the Working Party on the Suppression of Odours from Offensive and Selected Trades, Part 1, *Odours,* HMSO.

2.2   *Farm waste disposal: silage effluent,* Leaflet 87, Ministry of Agriculture, Fisheries and Food (1969).

2.3   *The story of animal glue,* Alfred Adam & Company.

2.4   L.J. Rubin, *The meal industry – progress and prospects,* Agricultural Institute of Canada.

2.5   R.E. Gardner, *The properties and applications of Bradford wool grease and derivatives,* City of Bradford By-products Department (1970).

2.6   E.V. Truter, *Wool wax,* Cleaver-Hume Press.

2.7   J.M. Oliphant, 'Dried poultry waste and intensive beef', *Agriculture* (December 1972).

2.8   J.G. Davis, *A dictionary of dairying,* Leonard Hill.

2.9   *Dairy effluents,* Ministry of Agriculture, Fisheries and Food, Department of Agriculture and Fisheries for Scotland, HMSO (1969).

2.10   *Rations for livestock,* Bulletin No. 48, Ministry of Agriculture, Fisheries and Food.

2.11   M.L. Windsor, *Fish meal,* Torry Research Station, Aberdeen (1969)

2.12   M.L. Windsor, *Fish Protein concentrate,* Torry
        Research Station, Aberdeen (1969).

2.13   *Condensed fish solubles,* Hull Fish Meal and Oil
        Company Limited.

2.14   *Food Investigation No. 13,* HMSO.

2.15   *Soils and manures for vegetables,* Bulletin No. 71,
        Ministry of Food, Fisheries and Agriculture.

2.16   C.G. Chumbley, *Advisory paper,* Ministry of Food,
        Fisheries and Agriculture. Now being revised.

2.17   *Agricultural use of sewage sludge,* Leaflet No. 47,
        Water Pollution Research Laboratory.

2.18   Report of the Working Party on Sewage Disposal,
        Taken for granted, HMSO (1972).

2.19   *Notes on water pollution: agricultural use of sewage
        sludges,* Leaflet No. 57, HMSO.

2.20   G. Hirst and B.M. Rock, *The Pudsey Project — an
        experiment in the direct re-use of sewage effluent for
        wool textile processing,* paper presented to Enpocon
        Conference (June 1973).

2.12   Firms, societies and other bodies consulted

W.H. Allen Sons & Company Limited

British Fur Trade Association

The British Wool Confederation

Canada Packers Limited

City of Bradford By-products Department

City of Manchester Rivers Department

Downton Tanning Company Limited

Glynwed Foundries Limited

Hull Fish Meal and Oil Company Limited

Iwel Engineering Limited

The Institute of Meat

Issac Spencer and Company (Aberdeen) Limited

Borough of Maidstone

D. Mason and Sons Limited

Ministry of Agriculture, Fisheries and Food

J.E. Olympitis

Borough of Pudsey

Scott Feather Company Limited

Stannard and Company (1969) Limited

Torry Research Station

Ross Poultry Limited

Unigate Limited

Chapter 3

VEGETABLE-BASED BY-PRODUCTS

The sources from which vegetation is procured have, in reality, only one origin, namely the supply given by nature. And since growing is a continuous process, supplies are never entirely exhausted unless the normal forms of plant reproduction are interfered with. Vegetation thus bestowed is consumed by man in every-increasing quantities and, to the natural processes, planned methods have been applied to give even greater production. Such intensification has increased the counter-products, the disposal of which has become a recurring occupation. These counter-products amount to harvest waste with little more than nuisance value. Indeed, in the case of forest waste much is just left to rot.

With regard to dual-purpose crops, a word of explanation is desirable. This term applies to two definite materials being obtained at harvesting from one plant, as in the case of the sugar beet. But it is often wrongly applied where processing a plant for a definite purpose liberates another material as a by-product. Neither does the term include under-sowing, which involves growing two classes of seed with different harvesting times on the same site.

The vegetable and fruit preserving industry is a special customer of the farmer. It is based on the necessity to keep, or delay, decomposition of such produce. From a cottage trade it grew to big business, invigorated by the acceptance of prepared foods, and later by the arrival of the so called 'convenience foods'. It would not be too much to claim that the blueprint of the trade is the supersession of traditional domestic cooking, or nearly so. En masse control in this manner enables useful waste constituents to be collected in sufficient quantity for them to be processed and marketed economically.

The examples of by-products from vegetable matter processing given in this chapter − by no means in order of importance − will serve to give an impression of what is currently going on, although they do not exhaust the subject.

3.1   Wood

The waste yield from the woodworking industry, from growing trees to finished end products, is said to amount to one third of the whole of the material involved. The fundamental difficulty in utilising this waste is, as in many other branches of industry in this country, the economics of collecting sufficient of uniform quality in one place to justify the installation of expensive processing plant.

Most of the wood used in this country is imported ready sawn, only about 6.5 per cent being home-grown. The early stages of waste production are, therefore, largely the concern of the countries involved.

Imported and home-grown timber is reduced in this country to the scantlings required by timber processing trades. The waste arising is composed of dust, shavings, off-cuts and rejects. Veneer and other specialist industries produce their own kinds of waste.

The sale of wood waste in the form of logs is age-old, but there have always been limitations because of seasonal changes. The decline in this trade is largely because of the wide-spread adoption of central heating and portable fires. Another contribution to this is the operation of the Clean Air Acts.

In some woodwork factories the wood waste produced has successfully been used in firing furnaces to raise steam and hot water. It is, however, not a procedure to adopt light-heartedly as it is expected to be supplemented by coal or oil. Such a typical coal-assisted scheme comprises a wood storage receptacle, a coal bunker and a transporter system. Suitable burning conditions are achieved by varying the fuel mixture. A typical arrangement is shown in Figure 3.1, although details are omitted. One refinement is a magne-

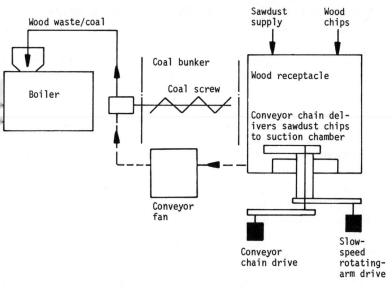

Figure 3.1 Steam-raising plant. *Courtesy:* GWB Boilers Limited

tic separator for dealing with incidental tramp iron. Steam so generated may be used for factory purposes or piped hot water.

The possibility of further extension of wood waste as a fuel for industrial use (and domestic heating as well) has been exploited in the form of briquettes, using a binder to produce density. Another approach has been compression to form cohesion without a binder. But it appears that until smoke emission is reduced little progress will be made in the general acceptance of briquettes.

The production of chipboard or wood-particle board arose out of the search for further uses of wood waste. Much of this is now manufactured out of forest thinnings and solid production waste. Some manufacturers of chipboard use varying amounts of waste but to ensure a high grade end product the material needs careful selection. This means the removal of sawdust, shavings and extraneous matter and involves the supplier in separation activities that can be difficult if the business is not completely integrated. The high standard demanded of acceptable waste and the handling charges involved make this outlet attractive only where favorable circumstances obtain.

A mixture of cement and sawdust has been successfully used in the manufacture of building blocks and slabs. The marketing of these seems to be satisfactory but not entirely progressive.

Wood waste as bedding for animals depends largely upon it being at one's disposal and the cost of alternatives. Shavings and sawdust is acceptable as litter for the chicken and broiler trades.

Wood waste compost, mulch and soil conditioners are attractive in some areas, and work is in progress to devise the best methods of operation.

Most of the exploratory work on using wood waste as a combined roughage medium and nutrient source for animals has been carried out in the USA. American estimates suggest that the *digestion quality* of treated sawdust is equal to that of hay. Future progress in this field will depend upon the material's ability to compete with other likely feeds.

Turning now to charcoal — despite the quantity of raw material — forest waste, off cuts etc. from saw mills — most of our requirements are imported. Manufacturing methods for this material have been rather static.

On the chemical aspect of wood waste, the following is quoted from a paper by Hall [3.1]:

'Wood contains 45-60 per cent cellulose and 17-35 per cent lignin, the remainder being made up of various hemicelluloses, small amounts of usually polyphenolic extractives and a very small amount of ash. In theory the potential uses for chemical processing are numerous and indeed many have been tried over the years. The basic difficulty is the usual one of the economics of getting sufficient material of uniform quality into one place to justify the cost of expensive equipment. In many cases there is competition with products which can be produced from coal or oil, supplies of which are still adequate in spite of earlier predictions. Wood could form the basis for producing alcohol, sugar,

yeast, plastics, oil, gas and many varied inorganic
chemicals but it is more economical to transport oil or
gas long distances than wood waste short ones'.

One of the most profitable outlets for wood waste is as a
material for pulping. The quality of the waste must be high
and its dimensions controlled to enable uniform processing.
Quoting Hall again:

'With current practice in this country chip production
is rarely feasible and the amount of waste processed
in this way is very small. In those countries where
primary and secondary conversion are major national
industries and where integrated processing plants are
possible because of the scale of operation, special
measures are taken to ensure that the maximum proportion of waste is produced as pulp chips. For
instance, the debarked logs are squared up by chippers
rather than saws producing sawdust and edge slabs.
It is debatable whether the industry in this country will
change in such a way as to make such measures feasible.
The number of pulp mills in the UK is very limited so
that, apart from in a few fortunate areas, transportation
difficulties would exist even if other circumstances were
favourable for pulp chip production. . . .'

There are numerous small and often local outlets for wood
waste, but none make more than a slight impact in reducing
the volume available. Dry sawdust is used as an absorbent
packing or as an absorbent sweeping compound. It is also
used in certain metal polishing operations and in the cleaning
and dressing of furs. In pulverised form it finds use as a filler
in some plastics, in linoleum and in polishing mixtures. There
is also a minor trade in moulded articles using wood particles
bonded with synethetic resin. Because of the increasing value
of vegetable turpentine its recovery from wood waste is now
attracting more consideration. Turpentine is a by-product
of the wood pulping process and recovery systems are now
being supplied, or contemplated, as an addition to paper-
making plants.

3.2   Cork

Cork comes from the bark of the oak. Many of its traditional uses are disappearing with the introduction of new materials and methods. A large portion of the cork trade is now concerned with problems involving flotation processes and moisture absorbtion. Granulated cork, using off-cuts and waste from various other processes, is used for insulation board, the process involved being to heat the material in its own resins. A more recent application of waste has arisen due to the sudden popularity of platform footware which has given a fresh impetus to the trade by compacting cork for soles.

3.3   Cotton

Waste arising from the spinning of cotton may be classified as either intentional waste or production residue.

Firstly, about 10 per cent waste is removed from the raw cotton during the early stages of spinning. These impurities, which are usually leaf, stalks and seed, are disposed of to waste materials merchants. Only the short fibre content is of any commercial value and this is sold off the processors of coarse yarns for use in the soft furnishing trade.

Much of the production waste is re-processed, more especially during the first stages of spinning. Waste made later is sold to merchants. Spun yarn is often used in making cotton rope and twine. Cloth off-cuts are sold to market traders for re-sale as remnants, or to waste merchants who sterilise them for use as industrial cleaners. Small pieces of cloth have an outlet to the papermaking trade.

Standards for faulty cloth created by weavers and finishers are usually set by customers, but in general such material is divided into two classes

1   Rolls of seconds (one fault in twenty yards): 25 to 100 yards.

2   Fent lengths: 1½ to 25 yards.

The usual outlet for these materials would be to fent merchants who specialise in disposal of this type of merchandise. About 50 per cent is exported, in the main to countries with lower standards than the UK, to the making-up trade and local bazaars. The remainder would find a home-market to remnant shops and market traders, and also to the make-up trade for 'Sale' items.

Generally speaking, the important constituents in the effluents occur in concentrations that are far to low to make recovery an economic proposition. They are usually residues of starch converted by enzymes, dilute caustic soda from the scouring of cotton, and dyestuffs removed during the washing of dyed and printed goods.

3.4 Rubber

The harvesting of rubber is of little consequence here, but the timber of the trees deserves a closer look. The rubber tree as a latex producer is constantly being improved and older trees are, therefore, regularly felled to replace them with higher yielding strains. Formerly, under the traditional process, the tree trunks were burnt in smokehouses to dry and smoke the rubber to produce smoke sheet. Rubber is now made by more modern methods involving drying in oil-fired hot air driers. There is, therefore, considerable interest in making full use of the wood for other purposes. Rubber wood is now used in the manufacture of furniture, teachests and fibre boarding and in the production of pulp for making paper and rayon. There is a comprehensive publication [3.2] entirely devoted to the utilisation of rubber wood.

From the seeds of the tree a useful oil, rather like linseed oil, is obtained and the residue from the process is suitable for cattle feed cake.

3.5 Straw

Figures computed in the year 1973 suggest that some 9.55

million tonnes of straw were produced in this country and of this approximately 3.55 million metric tonnes were surplus to requirements; it is estimated that a little over 60.7 per cent was used in agriculture, leaving a surplus of more than one third of the total.

A current procedure by farmers is the burning of stubble and some straw as it lies, but there are growing objections to this on the score of possible damage to the environment. The alternative is to turn the material back into the soil at ploughing time.

Changes in the present system of disposal would involve higher costs and labour difficulties the farming fraternity report. Even chopping, now technically easier, has problems.

The present usage of straw in manufacturing is in building board and insulation, and as a filler. There is also a small market for straw as a packaging material.

There are suggestions for the future. For instance, developments in feeding treated and untreated straw which could substitute for cereal and protein feeds offer potential. Another is straw pulp for paper-making, although suitable material may involve expensive separating.

3.6   Hops

The bine of the hop, after harvesting is either composted or burnt on site.

Spent hops arising from beer brewing have a certain value as a mulch-type fertiliser and are normally disposed of by the brewers through contractors to horticultural institutions. The spent materials are not usually processed and are just removed from the processing plant in wet form or slightly de-wetted.

3.7   Malt

A by-product of malt processing is spent grain which comprises husk and acrospire. The material, either in wet or dry form, is employed as an animal feed. Another by-product is culm — rootlets — which, difficult to handle and varying in quality, is sometimes used as a fertiliser.

3.8 Mustard

During the manufacture of mustard preparations two un-wanted substances require removal. They are the oil-bearing seeds and the husks; each is worth retrieving. From the former the oil is extracted by solvent recovery, the solvent being distilled for further use. The oil is refined to a com-mercially acceptable grade. The husks are marketed as an ingredient for animal feeds.

3.9 Tobacco

The stem or midrib of the tobacco leaf is unsuitable for manufacture into cigarettes or for packed tobacco. This is sometimes pulped and rolled into sheets for use in cigar manufacture.

3.10 Flax

This plant is grown mainly for its fibre or seeds, a different species being used in each case. Cultivation is predominately an overseas operation, the disposal of waste reapings and preliminary preparations, therefore, remain in the hands of the countries involved. In these countries a large amount of flax shive and fibre mixture derived from processing forms the basis of a lively business in manufacturing fibre-board for building operations.

So far as spinning in this country is concerned, any waste produced is mostly used with a lower grade of material ob-tained from the original spinning and weaving processes. Waste and small stuff, some of which is also suitable for paper production, is disposed of to merchants.

3.11 Hemp

Hemp is another plant that is cultivated abroad and imported into this country. The spinning of good quality hemp has not been carried on in the British Isles for several years, such yarns having been replaced either by flax or by synthetic fibres.

The bast removed during the course of yarn manufacture has been used in the manufacture of building board but this application has declined considerably. The industry is now so small as to render any use of by-products economically unviable.

Plumber's hemp, traditionally used in screw-jointing pipes, has almost entirely given way to plumber's flax (although the Trade Descriptions Act does not seem to have changed the original term).

3.12 Cocoa and coffee

The conversion of cocoa beans into cocoa and chocolate involves a series of separate steps during which the unwanted constituents — shell and germ (the radicle of the bean) — are removed by winnowing. By this operation the nib (roastings) are further processed and the shell and germ withdrawn from the production line.

So far little progress has been made in finding useful outlets for this waste beyond that of an auxiliary fuel for the manufacturer's own furnaces. A possibility being explored is that of a filler for plastics.

Powell and Harris [3.3] give a full explanation of the procedures leading to cocoa and chocolate products.

The residual material following coffee, coffee and chicory essence and proprietary substances is a mixture including spent coffee grounds. Many manufacturers have interested themselves in making this waste suitable for animal feeds, but until recently these investigations have not proved particularly fruitful. No cattle feeds from this source are likely on a commercial scale. The full economics of this project have so far not been published.

3.13 Jute

With the sole exception of cotton, more jute is consumed in the world today than any other textile fibre. Not being an indigenous plant, the United Kingdom's requirements are imported, mostly from India and Pakistan.

The material is the origin of an endless list of collateral products touching on many sectors of industry. Despite this there are really no by-products emanating from processing other than waste. Even the fats and oils in jute occur in insufficient quantity for successful commercial exploitation. With some types, however, a portion of the top fibre is cut off and marketed separately. This requires separate treatment to render it suitable for processing.

3.14 Sugar

In general, sugar is any sweet, soluble carbohydrate. In milk it is lactose; in grapes, glucose; and so on. The term is more generally applied to sucrose, the two most useful sources being sugar cane and sugar beet. Although methods of transforming raw materials into raw sugar are not fully described here, it is worth noting the by-products that are made available by these processes.

The accumulation of leaves following cutting down sugar cane is used directly as a fertiliser or indirectly as an animal feed. Leaf debris from feeding mixed with animal dung forms a high grade fertiliser. The harvested cane is crushed to extract the sugar juice and the crushed cane, known as 'bagasse', is used as an effective on-site fuel. Heat produced in this way is employed for generating electricity and the like. There is, however, a trend away from these kinds of utilisation where conventional fuels are available, especially now with growing outlets to the fibrous products trades. Miscellaneous uses of the material are in plastics, poultry litter and mulch, animal feeding, bagasse concrete and soil amendment. Bagasse as a useful by-product applies only in the cane-growing and sugar milling countries.

Raw sugar as presented for refinement is a brown moist

substance containing crystals surrounded by impure molasses. The steps in purification are the removal of the grosser impurities, then the insoluble impurities; next, the dissolved impurities, and finally the pure sugar is separated from the purified liquid. Briefly, the raw sugar is mixed with a warm syrup to soften the molasses and the mixture is centrifuged to spin off the softened molasses. Remnant syrup is washed off by water spray and the crystals go forward to the final stages of sugar manufacture. Spraying unavoidably carries away some sugar with the impurities and this is recovered by boiling, leaving a final pure molasses. As a by-product the material has a direct use as a fertiliser and as an animal feed, whilst the distilling industry uses it in the production of rum, ethanol, rectified spirits, etc.

The fermentation industries utilise molasses in the formation of vinegar and acetic acid, citric acid, lactic acid, glycerine and yeast. Miscellaneous uses are in the production of dextran and aconitic acid.

The only other waste for which extensive commercial outlets have yet to be developed is that of mud discharged by the filters. This is used as an animal feed and a fertiliser. The mud contains a wax that coated the sugar cane whilst growing. This is a valuable by-product, but the evidence suggests that its separation from the mud is not always viable. It was first isolated by Avequin in 1841 and subsequent ventures in various parts of the world at later dates came to an end. In 1958 there were various plants isolating the wax, but it seems that only one plant in Cuba is still operating.

Since the by-products from the sugar beet industry are of home origin they are of significant interest to the local farming fraternity. Indeed, the two factions — grower and processor — depend on each other. At harvest time the beet tops and their foliage are cut off; some are ploughed back into the soil, while others are used as cattle feed.

The preparatory operations in the manufacture of beet sugar are washing and slicing. The material then goes through several stages of development, the main ones being extraction of the sugar-laden juice from the slices, refining the sugar, removal of surplus water and crystallising the sugar for marketing. The spent slices are pressed to reduce their moisture content and to the mass is added molasses to form dried sugar beet pulp. This material is used in making-up livestock rations. Beet sugar requires quicklime at one stage in production. It re-appears as lime with other production waste which, after processing, is marketed for agricultural purposes. It is rich in calcium and also contains phosphorus, magnesium and nitrogen — all plant nutrient elements. The main value of this material is for neutralising soil acidity. The pure molasses arising, if not used with pulp, are employed for the same purposes as with cane sugar.

3.15   Cider

Cider production in this country is a seasonal operation, the apple crop being raised by farmers with mixed farms as opposed to specialists fruit growers. During harvest time the apples are shaken from the trees and delivered immediately to the cider makers. They are graded, washed, polished and reduced to pulp. The apple pulp is then expressed for the separation of juice which eventually forms the cider. The resulting pomace, being a ready-to-hand source of pectin, promotes even factory performances throughout the year.

Whilst all pectin was originally sold in liquid form, this form is now only used in any quantity by jam manufacturers in the UK. In all other situations powdered pectin is used. Although jam making still accounts for the major part of the world's consumption of pectin, increasing quantities are now being used in the field of confectionery and dessert products, this being stimulated by the availability of certain chemically modified pectins.

The spent pomace is dried and sold for animal feed, whilst apple sludge has been used for distribution over land by an irrigation process.

3.16   Artichokes

The lower parts of this plant with a thistle-like foliage and bearing terminal flower-heads is used as a food.

Harvest waste is suitable for composting, otherwise further uses are minimal. A possible future by-product is the heart which, according to experiments at Yale University, USA, suggest they can be used for sweetening water. Efforts are now being made to isolate the substance that causes this taste delusion. The aim is to avoid the need for artificial sweetening substances in food and drink.

3.17   Apricots

This tree-fruit is imported into this country and much used in the fruit preserving trade. The nut, unwanted in fruit preserves, is crushed and the kernel separated. From this the oil is expressed and refined for use by the fish-canning industry.

3.18   Lentils

Lentils are the seeds of a leguminous plant grown in various parts of the world. The considerable harvest waste is suitable for feeding to animals. It is usually mixed into compounds by manufacturers and distributed via merchants to farmers.

3.19   Vegetable oils and fats

All plants contain oil, but few in sufficient quantity to justify extraction on commercial scales. Among those that are so viable are the coconut, palm fruit and kernel, groundnut, sesame, sunflower, soya, cotton, teaseed and rapeseed. Whilst many of these are processed in the countries of their origin, such as Russia and China, vast quantities are sent to other areas where markets are concentrated. By crushing the cellular structure of the plant the oil content is liberated. Milling methods vary to suit different plant characteristics, but the general pattern of procedure is the same. Figure 3.2 is typical of the process for groundnuts, the programme being

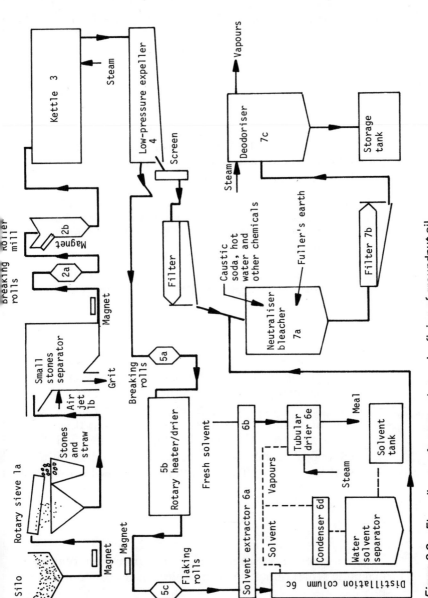

Figure 3.2 Flow diagram for extraction and refining of groundnut oil

summarised as:

1   *Cleaning* – Seed from the silo is cleaned by (a) rotary
sieve and (b) an air current.

2   *Breaking up* – The seed is reduced to a course meal
between (a) heavy fluted rollers and (b) plain rollers.

3   *Heating* – In the kettle, rotating arms stir the broken
seed as it gravitates through steam-heated pans. Heat
expands the oil and moisture in the seed cells, which
burst and release their oil.

4   *Expelling crude oil* – The expeller is a two-delivery
type mincer with a parallel cylindrical housing containing a tapered screw. This squeezes out the crude oil from
the solid matter.

5   *Preparing for solvent extraction* – The seed, now in
cake form, is (a) again broken up, (b) dried and (c)
flaked, before reaching the solvent extractor for the
removal of more oil.

6   *Solvent extraction* – (a) Jets spray solvent on the flakes
as they are carried on a moving belt of fine wire mesh.
(b) The solvent/oil solution falls to trays set on a des-
cending level, and as it overflows from one tray to
another it is pumped up to the next jet. So the concentration of oil in the solvent increases until it flows from
the lowest tray. (c) The solvent is evaporated and distilled, and (d) condensed for re-use and the crude oil is
sent for refining. The remaining flaked seed is used as a
meal after the residue of solvent has been dried and
condensed.

    The accidental occurrence of tramp iron is overcome by
magnetic separation at strategic points in the production
line.
    In the case of de-shelled groundnut, milling can remove
about 44 per cent of the weight of the seed, which appears
as oil, but 46 per cent can be extracted as oil by soaking

partly expelled seed in light spirit. The solid 54 per cent which is left includes less than 1 per cent of oil. This solid material is a valuable feeding stuff for animals. Some other varities of seed are also so used.

Some classes of harvest waste, such as leaves and stalks, are also suitable for fodder, and in a few cases similar matter may be economically exploited. For instance, the flowers of the sunflower yield a yellow dye, while in some countries the stems are burnt and high-grade potash salts extracted.

The significant place in industry held by the coconut is largely because it is a source of vegetable oil. But its shell also has value, and the paring (the green skin) also contains an oil. The production of parings oil has become quite important in Sri Lanka (Ceylon). Which of the two parts, nut and shell, is considered the main product leaves the other as the by-product. A fully integrated factory processes both.

Coconut shell as a fuel, apart from its calorific value, has the advantage of waste reduction.

Much of the active carbon produced today is made from coconut shell activated by selective oxidation with steam at high temperatures. Something like three-quarters of the shell content is volatile matter and moisture, which is mostly removed at the source to reduce transport costs. Houghton and Wildman [3.4] point out that the cellulosic structure of shell determines the end product, which at normal yields of 30-40 per cent on the carbonised basis is a material with a large internal surface consisting of pores and capillaries of molecular dimensions. Normally the ash content is low. Manufacturing methods are not discussed here, it being sufficient to say that a variety of grades and qualities of activated carbons result. Active carbon has the ability of absorbing even traces of either unwanted or valuable liquids and gases. This is the chief area of application, where it plays an important part in solvent recovery processes, water and effluent treatment, and the clarification of flue gas before discharge to the atmosphere.

The kernel is mainly ground or desiccated and used in many cooking processes. The 'milk' is sometimes used as a beverage in its natural state, fermented and made into palm wines and vinegars, or used to produce a form of sugar. Oil obtained from the dried kernel is employed as an ingredient in the production of margarine and, to a much lesser extent, soap, and it is valuable in vegetarian diets. The fibres of the shell are still the material of coconut matting, although synthetic fibres are now more favoured for this particular application.

3.20 Essential oils

These are volatile oils derived from the leaves, flowers, fruits, stems, woods, wood mosses and roots of certain plants. They form the basis for perfume making, distillation and solvent extraction being prominent in processing. By far the greatest volume of waste is spent fibrous matter which may be used as an auxiliary fuel, as in the case of lavender in this country. The solvent extraction process results in a wax. This consists of the final floral essence mixed with beeswax. The wax, after refining, is a by-product of the process.

3.21 Oats

Of the total amount of oats processed in the United Kingdom, about 50 per cent is for human consumption in the form of flour, flakes and meal, the rest being used for animal feeds. Milling follows generally the same pattern as for many other grains. The offal from the process is mostly husk, the covering of the seed. By and large this is used to form oat-meal by-products, which, by the Fertiliser and Feed Stuffs Act, must conform to a given specification. The mild chafing characteristic of the shell renders it suitable as an agent for polishing and for de-carbonising internal combustion engines. Further quantities are used by linoleum manufacturers and, when finely ground, as a dispersible powder for carrying insecticides.

3.22   Wheat

The wheat used in this country is derived both from overseas sources and from our own wheat-growing areas. Much of the home-grown wheat is used to produce biscuits and cake flour and flour for the housewife. The remainder is blended with imported wheat to make flour for the nation's bread.

The following is typical of the milling operations involved:

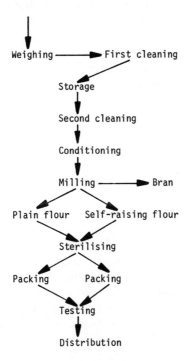

The by-product, mostly bran, arising from milling is almost entirely utilised for animal feeds. Most is sold to manufacturers of animal-feeding compounds for incorporation into pelleted foods for pigs, poultry and other classes of livestock. Some millers, however, do conduct a separated trade direct to farmers, although this is declining. The large flour millers often engage in this specialised production of animal feeding-stuffs. More profitable is the separation of bran for horse-feeding, but this is undertaken by only a few millers. Millers by-products fluctuate in price, depending on the current availability of animal-feeding stuffs from other sources, but the price they command is less than the price for flour itself.

3.23   Maize

Much of the corn in America is commercial hybrid corn — as distinguished from sweet corn and popcorn, which are specialty crops. It is estimated that some three-quarters of the annual crop is fed to livestock and poultry. A significant amount of the remainder is purchased by millers and, to a less degree, by distillers.

The long and varied list of by-products and collateral products arising from commercial hydrid corn have no real parallel in the milling of wheat for flour. Refining operations vary to some extent between mills, the following short description being fairly typical. Corn deliveries are cleaned, stored and cleaned again before going to steep tanks in warm water. The steepwater is drawn off, and the softened kernels go to degerminating mills and on to separators where the oil-laden germs are removed and oil extracted. The remaining slurry of starch, gluten and hulls is ground, then passed through reels and shaken to remove the hulls. Starch and gluten are then separated in centrifuges. Free of corn's other elements, the starch is washed, dried and prepared for sale as starch and dextrin, or converted into syrup and extrose. Maize products and by-products have, in recent years, permeated industry at such a pace that it is difficult to keep up with them. The main headings under which they fall are given in Figure 3.3

3.24   The vegetable and fruit preserving industry

The divisions of this industry could be stated to be canning, the manufacture of jams, savoury flavours, sauces, ketchups and salad creams, the drying of fruit and the brewing of malt vinegar.

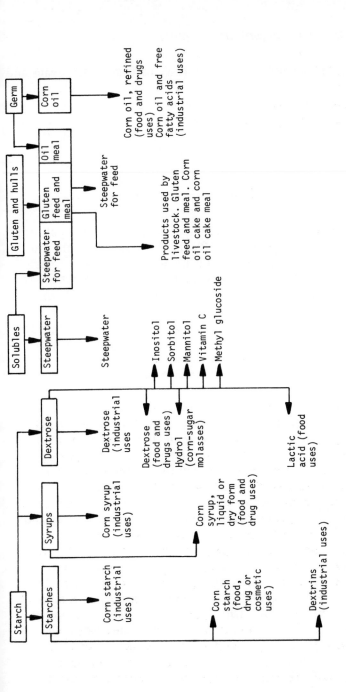

Figure 3.3 By-products and collateral products from maize

Vegetable canners can fresh peas, carrots, broad beans and new potatoes, etc., in their respective harvesting seasons, and process peas, baked beans and butter beans at all times. From these activities two forms of waste arise: (a) an effluent containing solids in suspension and (b) solid matter. The former is chiefly peelings (where applicable) and washings (split peas and beans, pea pods, stalks and small rejects). These solids are comminuted to comply with the particle size standard laid down by the drainage authority concerned before the effluent is discharged into sewers. Some canners screen out these solids, which are suitable as an animal feed, before being so discharged. By far the majority of the waste is in the latter category, and is made up of physical rejects and spillage. Physical rejects are such items as carrot tops and blemished vegetables that are removed by hand from inspection belts, whilst spillage is litter accumulated during handling. This waste is collected and sold at nominal prices to local pig farmers for ultimate usage as pig swill. Whilst the normal expansion of the industry is expected to increase waste this is partly offset by pre-cleaning and washing of the vegetables coming into the factories, e.g. fresh peas and broad beans are vined, cleansed, washed and cooled on farms, and carrots are washed and sometimes topped before delivery.

For about three weeks during the harvest season quantities of soft fruit are canned, thus concentrating a temporary additional problem. Rejected fruit is separated out and joins other vegetable swill. Small and large damaged berries are chemically preserved in containers for delivery to jam-makers.

Waste from jam making is chiefly fruit stones, seeds and skins. Citrus fruit peel, whilst separately processed, is an integral part of the fruit and is treated as such, performing a valuable function in the finished product. Citrus peel is also used in the production of soft drinks. Rejected fruit forms a relatively small proportion of the waste because raw material purchases are strictly controlled from contracted sources. A small amount is, of course, inevitable and from this the best is usually pulped for use as a fruit flavour additive in products such as ice-cream. The remainder, together with the skins and plant haulm, can be disposed of as animal feed. Stones and seeds are sometimes ground and used for metal polishing, but usually because of the irregular supply they are either dumped or incinerated.

The production of sauces, ketchups, salad creams, etc., forms one division of the preserving industry. The major raw materials used include malt and barley for vinegar, vegetable oil for salad creams, tomatoes for ketchups, dried fruits for brown sauces and frozen fruits or fruit concentrates for sweet dessert sauces. In addition there are many minor ingredients, e.g. onions, garlic, spices and starch thickeners.

The quantity of waste from processing is relatively small, arising mostly from vinegar and brown sauce production. A small quantity of raspberry seed can arise from sweet raspberry sauce, but from the other dessert sauces already processed fruit pulps or concentrates are used. The tomatoes used in ketchups are in the form of concentrated pastes, all waste having been removed in the concentrating factories of the suppliers.

In the case of vinegars, a fair amount of waste spent grain is produced and this goes for cattle feed. A small quantity of yeast by-product is also sold for conversion to yeast extract.

From brown sauces, those fruits which are not purchased as ready-processed concentrates are dates, raisins and tamarinds. These are simmered in vinegar and the pulps sieved in an early stage of the processing. The sievings consist largely of stones and fibre, which are apparently useless for any purpose manufacturers have been able to suggest. At a later stage, a finer seiving removes materials such as fruit and onion skins, small raisin stones, coarse spice particles and the like. Manufacturers have received a number of enquiries about the use of these fine sievings, but none have come to satisfactory conclusions. The difficulty appears to be the acid and highly-spiced nature of the materials which makes them unsuitable for animal food. They are also resistant to decay. Onion peelings are included in this material.

At the present time this and the associated industries are at pains to find lucrative outlets for their waste. Many avenues have been explored without success.

Malt vinegar produces large quantities of usuable by-product in the form of spent grain which comes from the whole malt and barley used in its manufacture. This material contain cellulose and sugars and is sold for pig food. Under the old methods for malt vinegar manufacture, quantities of yeast were produced during the fermentation stages and this was sold to manufacturers of yeast products. Under modern methods little yeast body is produced and therefore no saleable amount is produced.

Imported fruit for the preserving industry is scrutinised for deterioration and damage. Deteriorated fruit is rejected and destroyed, and damaged fruit is sold cheaply for animal feeds.

3.25   References

3.1   G.S. Hall, Timber Research and Development Association, 'Utilisation of wood waste', *Chemistry and Industry*, pp 781-84 (1971).

3.2   *Rubber Research Institute of Ceylon Bulletin*, **5** (3-4), (September — December 1970).

3.3   B.D. Powell and T.L. Harris, 'Chocolate and cocoa, *Encyclopaedia of chemical technology*, Volume 5, pp 363-402.

3.4   F.R. Houghton and J. Wildman, 'Manufacture and uses of active carbon', *Chemical and Process Engineering* (May 1971).

3.26 Firms and other bodies consulted

John Baynes Limited (Baynes & Dixon)

Baxter Brothers and Company

J. Bibby Food Products Limited

Bovril Limited

British Sugar Corporation

Brooke Bond Oxo Limited

H.P. Bulmer Limited

Cadbury Schweppes Foods Limited

Co-operative Wholesale Society Limited

Corn Refiners Association Incorporated

Courtin & Warner Limited

English Seafoods

The Flour Advisory Bureau

Malayan Rubber Fund Board

Morning Foods Limited

Nestle Company Limited

R. Paterson and Sons Limited

Frank Pugh and Company Limited

Rickett & Coleman

Sidlaw Industries Limited

Smedley-HP Foods Limited

South Mills (Flax) Limited

Swel Foods Limited

Tate and Lyle Refineries Limited

Timber Research and Development Association

Tootal Limited

The Trent Yeast Extract Company Limited

Tricity European Sales Limited

Tunnel Refineries Limited

UML Limited

Unalco Limited

Unilever Educational Publications

Vagan and Company Limited

Chapter 4

MINERAL-BASED
BY-PRODUCTS

A mineral is any substance (generally inorganic) occuring naturally within the earth or on its surface. Our mineral wealth, shaped and formed million of years ago, has been sequestered by man in many forms. Industry continues to consume this irreplaceable harvest at an ever-increasing rate. The rate has become so high of late that for many minerals present sources are likely to have been completely used up in the foreseeable future. Even at the present rates of consumption, oil and natural gas may last only another 100 years and coal 300 to 400 years. An authority [4.1] on metals states that a few years ago it could be said that more of some metals had been mined in the last forty years (in some cases, irrecoverably used) than in the whole of previous history; in 1972 the figure is beginning to look more like twenty-five years. The curves of production of some metals against time given in Figure 4.1 are typical of what is happening; their form is also exponential when smoothed. The following has been culled from the paper:

> 'This brief survey suggests that no supply problem exists
> for aluminium and iron for many generations; that
> copper and nickel can be maintained for at least fifty
> years and maybe longer, but at a rising price, and that
> the same probably applies to platinum. For lead, tung-
> sten, cadmium, germanium, silver and particularly mer-
> cury, a real question will be beginning to emerge about
> new sources of these non-renewable resources by the
> end of the century.'

Recent attention has been centred on Britain as a source of non-ferrous metals as supplies elsewhere diminish. Large-scale

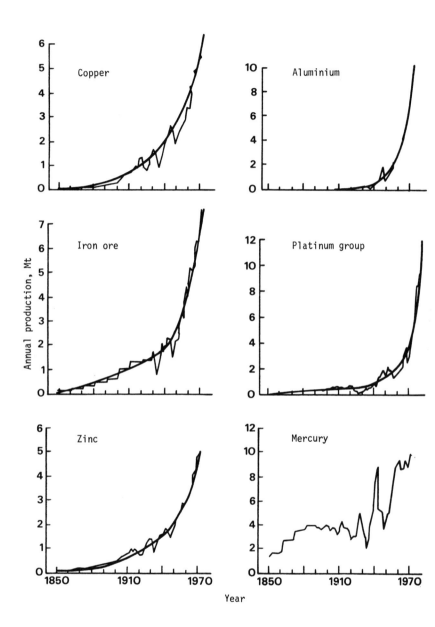

Figure 4.1 Production of metals, 1850-1970, with actual and smoothed curves. Prepared by the Mineral Resources Division of the Institute of Geological Sciences

operations in this country might be needed to win a few tons of non-ferrous ore. But, as the economists point out, with imports running at over £600 million a year, any home-produced contribution could do wonders for the balance of payments.

Strong growth and lessening available resources are appearing in many other wings of mineral winning.

We have to ask how long the known resources will suffice in a situation where requirements are doubling every decade. It is not difficult to single out the need for the discovery of fresh mineral deposits to ensure the future. In 1854 Lapworth [4.2] said 'As we are unable to penetrate more than a few thousands of feet into the substance of the globe, the labours of the geologist are necessarily confined to its solid and accessible exterior'. This limitation is largely removed and today the deepest boring is 30,050 feet for a gas borehole in Oklahoma. Already geophysical surveys have disclosed new mineral reserves and these are so enormous that there is no immediate fear of depletion. So it seems that the problem in the future is likely to be one of accessibility rather than one of shortage.

Another factor that cannot be thrust aside is the constant clash between the exploiters and the champions of the environment. Both put forward arguments that are difficult to rebut. More and more legislation is bound to bring new troubles to overcome.

The tendency for manufacturing activities to be concentrated into decreasing numbers of large units is followed in the mineral industry. This is particularly the case where there are ample indications that there are sufficient reserves to last a long time, so that it is viable to integrate winning and processing. Such is the case in the slate, gravel and dolomite industries. In other instances, e.g. oil, refining is not done at the place of extraction. A concentrated industry is seemingly accountable for a more agreeable by-product bonus than a splintered one. Against this it could be said that processing materials where markets exist is equally important.
important.

Lastly, there is the incidence of synthetics. Currently there is a substantial increase in the development and use of these in place of many traditional mineral substances or their derivatives. But, at best, substitution can only be a partial solution to the resources dilemma. This does not mean that recycling is being neglected. Far from it. It means the need to seek out new and novel reclamation techniques that are likely to show promising lines of attack.

One such unusual reclamation technique, currently being investigated, avoids remelting. By this method machine swarf has been reconstituted to form rod and forgings. The mechanical properties obtained are claimed to be high and, from initial investigations it appears that the process could be made commercially viable. The process involved is explained by Dower [4.3].

4.1   Rocks

The materials falling under this heading are chalk, coal, slate, granite and the like. Some of these are hard and compact, others soft and loose. But no matter what may be their degree of consolidation, all are known to the geologist as rocks.

4.2   Ore

Ore is a mineral body from which metal is extracted; the compound of a metal and some other substance, such as oxygen, sulphur, or carbon, called its mineraliser. Some metals, however, occur naturally in an almost pure state, e.g. gold and lead.

4.3   Mineral industries waste

The sum total of waste originating from solid mineral winning is unavoidably enormous. Apart from 'back-filling' where possible, it amounts to a by-product in the wrong place. Re-use is bound to involve heavy transport charges and financial viability alone seems remote. The one thing that is quite clear is that customary methods of disposal will have to be foresaken as environmental, ecological and anti-pollution legislation increases.

4.4 Coal

Coal is a material formed by the decomposition and compression of vegetable matter over many thousands of years. The formation may have taken place in the sequence; peat, lignite, bituminous coal, anthracite.

When the National Coal Board was formed in 1947 it inherited the legacy of 200 years of dereliction — an era of widespread ecological neglect. All this despoilation could not be remedied in 25 years of nationalisation, but considerable progress has been made.

The Board claims that so far 10 million tons of waste materials (from pitheads and slag heaps) have been sold for use in road works and ancillary schemes. With the continuing emphasis on road expansion there is the opportunity to clear and use much of the 2,000 million tons of shale contained in 2,000 spoil heaps within the Board's jurisdiction.

While red shale from the old burnt heaps has been used for many years, it has now been proved that the far more common black shale can also be used on a vast scale.

Another benefit is that contractors do not always need to dig so many 'burrow pits' along the sides of roads from which they obtain filling material, afterwards abandoning them.

Marketing shale is a means of removing pit-heaps, and a pattern of co-operation with local authorities also assists in landscape reclamation. Landscaping has also been carried out on more than 100 Coal Board sites where the land has been sold or leased, transferring to local authorities about 4,000 former derelict acres.

Acting independently, the Board has regraded, soiled and grassed scores of defunct heaps. This treatment included planting wild roses at West Cramlington, Northumberland, where the hip harvest was sold as the basis of a vitamin C health preparation.

Research continues to provide further uses for shale. For example at Pegwell Bay in Kent, land was reclaimed and a Hoverport built on a foundation of some ¼ million tons of colliery shale.

4.4.1   Coal processing

When coal is heated in the presence of air it burns to produce one single useful item — heat. When it is heated in the absence of air it decomposes into gas, coal tar, a watery liquor and a solid residue. This process of carbonisation is used by three industries:

1   *The gas industry* — where the main object is gas making.

2   *The steel industry* — where the main object is coke for feeding to blast furnaces.

3   *Low temperature carbonisation industry* — where the main object is smokeless fuel.

Murdock (1754-1839) is usually attributed to having been the first to produce gas from coal in a retort, although he did not extend his experiments beyond that. This was left to Perking, who, in 1857, set up a factory to manufacture synthetic aniline dyes from coal tar products. This was the beginning of the aniline dye industry.

In the United Kingdom at the present time the production of gas from petroleum naphtha feedstocks, and more recently the growing use of sea gas, has resulted in a steady falling off in gas works carbonising which will eventually cease altogether. Nevertheless, the industry still manufacturers some gas by the original method.

Before nationalisation of the coal and gas industries, coke ovens and by-product recovery plants were operated by most of the independent gas companies and local authority gas departments. Refining of crude tar and benzole was generally carried out by a number of co-operative tar distillers. The major changes in the industry have been brought about first by nationalisation of the major industries operating coke oven plants and secondly, by the decline in the gas works carbonising industry. At present coke ovens are operated by the British Steel Corporation and the National Coal Board, although there are a few independent operators. Low temperature carbonisation in retorts is carried out by two major companies and there is a small amount of carbonisation at gas works, mainly in retorts. Refining of crude tar is carried out by BSC and NCB, there being only a few independant distillers left. Refining of crude benzole is carried out only by the BSC and by Staveley Chemicals Limited, which is jointly owned by the BSC and NCB.

The coke ovens are arranged to make a coke suitable for feeding to blast furnaces used in steel making. Others are designed to form blended coke used in producing castings and other engineering materials, whilst facilities for the domestic market are catered for.

Advances in blast furnace design and the introduction of fuel injection techniques, have resulted in a decline in consumption of coke per ton of iron produced and this is expected to decrease further during the next decade. In the UK however the major output of crude steel has been from open earth furnaces and the industry is now changing over to production from basic oxygen furnaces with a smaller amount from electric arc furnaces. The basic oxygen furnace requires a higher percentage of pig iron in the feed than an open hearth furnace. This means two conflicting trends. The ratio of pig iron per ton of crude steel is increasing, but the coke requirement per ton of pig iron is decreasing. These are tending to balance out, but the natural growth in the total output of steel means that the overall requirement for coke is on the increase. There is one further complicating factor and that is the shortage of supply in this country of suitable grades of coking coal, which will inevitably result in the importation of much larger tonnages of coal. The type of coal imported will almost certainly have a lower volatile content than most of the indigenous coals, which will result in a lower yield of by-products than at the moment. This all seems terribly complicated, but as a result of this it is still expected that the output of by-products from the steel industry will increase over the next decade.

The output from the NCB and independent operators will depend to a large extent on the markets for solid fuels and these will almost inevitably take a much smaller share of the energy market, both for domestic and industrial purposes.

4.4.2   The coking process

Coke is the substance left when the volatile parts have been distilled from coal by heating.

Modern by-products coke ovens are arranged in batteries of up to sixty units, each unit being married up to a heating flue and associated pipework. The units are heated by combustion gases passing through the flue, the units being maintained at a temperature of about 1300°C. The fuel for this purpose may be coke oven gas, blast furnace gas, or a combination of both. The calorific value of the former is 560 BThU and the latter 80 BThU. Therefore, two extremely varying gases are sometimes mixed and it is in this form as a mixed gas that the gases are used as a fuel.

4.4.3   By-product recovery

A number of methods of recovery are employed, the following being typical.

The volatiles are piped from the ovens to a liquor spray where most of the tar and a large amount of ammoniacal liquor is removed by condensation. This is followed by further cooling which removes additional tar and ammonia liquor which are separated by decantation. The tar is then pumped to a heated storage tank until excess water has had time to settle out when it is fed to a tar distillation plant. The ammonia liquor is transferred to a still where ammonia is removed by treatment with steam and lime. This joins the ammonia-laden gas stream which passes through a solution of sulphuric acid in an absorber. The ammonia combines with the acid to form ammonium sulphate which is recovered in crystalline form.

The ammonia-free gas is now treated to remove benzole.

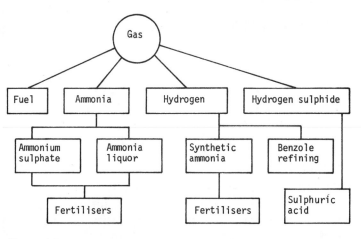

Figure 4.2 Major markets for coke oven chemicals: gas

After further cooling the gas is brought into contact with a
counter-current of absorbing oil which absorbs most of the
benzole in the gas. The benzole is then separated from the
oil by stripping and after separation from the water the
crude benzole is delivered to a benzole refinery.

The coke oven gas has now been stripped of its tar,
ammonia and benzole but a small content of hydrogen sulphide remains. Various gas desulphurisation processes are in
operation and may be linked with sulphuric acid production.

The major part of gas produced previously went into the
National Grid, including most of the gas produced by the
NCB. The introduction of natural gas has removed this
market and the major use is now as a fuel at steel works
and some other industrial undertakings. Hydrogen sulphide
is removed because it was not tolerable in town gas, because
of technical requirements at steel works, and also to reduce
atmospheric pollution. The chief process used in this country produces sulphur as a by-product and the use of this sulphur in the production of sulphuric acid is incidental (in
many cases the quantity is relatively small).

Figure 4.2 shows the major markets for coke oven gas
products.

4.4.4 Crude tar distillation

Crude tar is first subjected to distillation by which it is
fractionated into a number of distillate oils and a residual
base tar or pitch. As coke oven tars are relatively rich in
naphthalene and anthracene, primary distillation is chiefly
designed to yield fractions from which these materials can
be recovered. The usual distillation procedure in the UK is
by continuous operation. Water and volatiles are separated
from the crude and dehydrated tars, and the tar vapours
are separated into a series of fractions. These fractions are
refined or processed to obtain naphthalene, anthracene,
phenols, cresols, pyridines, quinolines, creosote and pitch.

A typical range of fractions and the chemicals extracted
from them are shown in Figure 4.3.

4.4.5 Refining of crude benzole

Crude benzole recovered from coke oven gas is composed
largely of the aromatic hydrocarbons, benzene, toluene and
xylene, with unsaturated hydrocarbons and a small amount
of paraffinic material. The light-oil fraction from coal tar
contains these same compounds together with trimethyl
benzenes and hydrindene and is usually treated with the
crude benzole.

Extraction of the pure hydrocarbons from the crude
benzole involves firstly a preliminary distillation to remove
low-boiling-point compounds, chiefly carbon disulphide.
The treated benzole is then further treated to remove sulphur
compounds and unsaturated hydrocarbons although as shown
in the flow chart further fractionation may be carried out
and only a selected 'BTX' fraction treated in this way. Subsequent fractionation separates benzene, toluene and a
mixture of xylene isomers. Further treatment of fractions
may be carried out, for example solvent extraction to
remove paraffins. The chief processes used for removal of
sulphur compounds and unsaturated hydrocarbons are acid
washing and hydrorefining. Washing with strong sulphuric
acid converts thiophenes and alkyl thiophenes to sulphonic

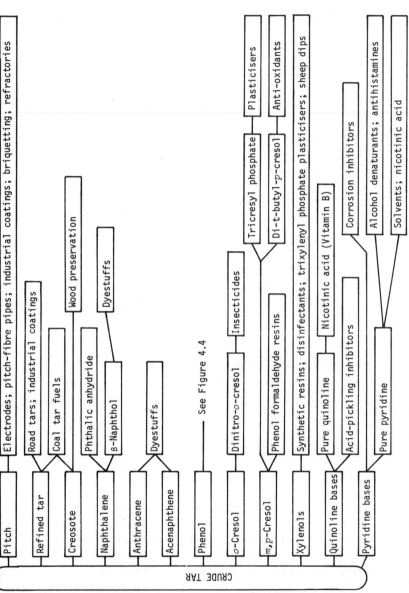

Figure 4.3 Major markets for coke oven chemicals: crude tar

acids and polymerises unsaturated compounds. Catalytic desulphurisation by hydrorefining involves selective hydrogenation of sulphur compounds, and saturation of unsaturated hydrocarbons.

The heavy end or naphtha fraction of crude benzole contains the unsaturated hydrocarbons, indene and coumarone, which are polymerised in solution to yield coumarone indene resin.

The major markets for crude benzole from coke ovens are given in Figure 4.4 whilst Figure 4.5 is a typical flow diagram of chemicals from coal carbonisation in general.

4.4.6   Treatment and recovery of fine coal

When coal is cleaned by wet processes the fine products are usually first de-watered on screens with ½mm aperatures. The result is a large volume of water containing some 10 per cent. solids by weight from which fine coal is recovered for mixing into the saleable products and the water for re-use.

At this stage the coal solids in the slurry are in the 'raw' state in the sense that most cleaning processes are relatively ineffective on material below ½mm in size. However, with most industrial fuels thickening of the slurry followed by filtration yields a filter cake that can be mixed with clean 12.5mm − 0 coal without serious effect.

In the case of coking coal however, the fine coal has to be recovered in a clean state and this is done by froth flotation. The reasons for this concern not only ash reduction but also the fact that certain of the coal constituents vital to the coking process concentrate naturally in the fine sizes. The clean coal froth is filtered and mixed with the cleaned coal leaving the tailings to be dealt with separately. Tailings are essentially fine particles of carbonaceous shale with an ash content of 70 per cent or more suspended in water. The solids concentration is rarely higher than 4 per cent by weight so that water recovery is essential. Thickening in large diameter thickeners yields clarified water and tailings 'slurry' containing about 35 per cent solids by weight. These solids, which form the only true by-product of fine

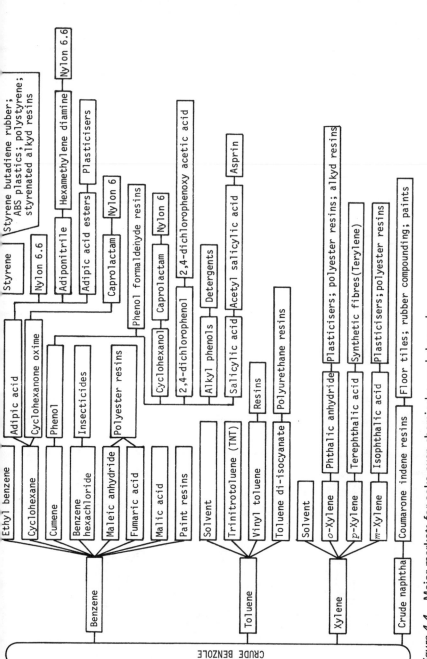

Figure 4.4 Major markets for coke oven chemicals: crude benzole

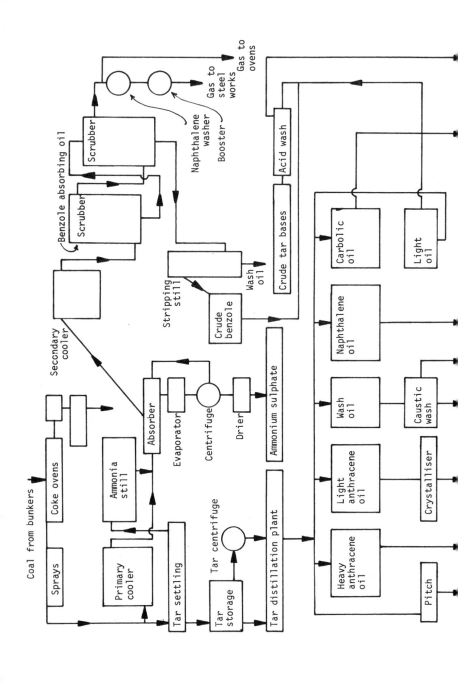

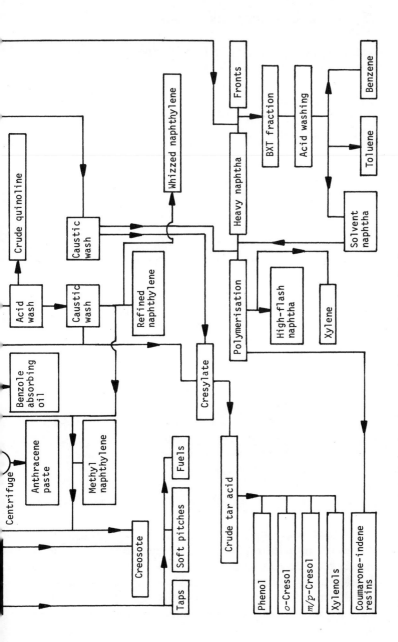

Figure 4.5 Typical flow chart for chemicals from coal carbonisation

coal cleaning operations, may be further de-watered in pressure filters or pumped to lagoons.

Of necessity this description is in general terms and in practice there are a number of variations. There are difference in practice between the UK and the USA [4.4] but the theory remains the same of course.

4.4.7   Pulverised fuel ash

Many Central Electricity Generating Board power stations burn pulverised coal as a fuel and this results in a powdery ash as a by-product. The ash may be removed in tanker wagons as a dry powder, as a water-sprayed material (to reduce dust hazards) or transported as a slurry.

The current annual production of pulverised fuel ash is over five million tons and is expected to rise to fifteen million tons by the early 1970s. Exploratory research into uses for this material has been carried out by the CEGB for several years. The most promising uses are probably those connected with the building trade and an increasing usage by various processing industries. The bulk of the material, however, still finds its way on to the land.

Much of the ash deposited on land has been spread in the locality of the power station producing it. This, of course, is to minimise transport cost. Other amounts have been used to raise the level of low-lying land liable to flooding. One such site is along the Dee estuary where new land has been created by means of the ash from Connah's Quay power station.

There are occasions when pulverised fuel ash has been spread on agricultural land with beneficial results. It is hardly reckless to suggest that this is a procedure likely to develop as convenient dumping sites become scarce and the effects of ash on plant growth become better known. Detailed information on this subject has been published elsewhere [4.5].

4.5 Oil

Crude oil recovered from natural reservoirs and oil-bearing strata deep underground is the raw material of the oil industry. Sometimes this oil is found with natural gas, a valuable product in itself. Often, however, natural gas is found alone – 'unassociated gas' – as it is called. When the gas contains hydrocarbon liquids or condensates these can be a useful additional source of oil refinery feedstock.

Oil refineries produce a large range of petroleum products which are not related to liquid fuel. As examples may be mentioned lubricants, greases, bitumens and chemical feed-stocks such as special fixed boiling-point spirits, medicinal and technical white oils, waxes and petroleum jellies. These substances and others once considered by-products are no longer claimed as such. Indeed, the term has almost disappeared from the oil industries vocabulary.

Refinery methods vary. The following procedures together with Figure 4.6 being typical, refer to the refinery at Fawley.

4.5.1 Primary distillation

This is a process of fractional distillation. Crude oil is first heated and the vapours produced are then condensed in a fractionating column to produce fractions of different boiling ranges. These raw fractions are subsequently further refined or purified.

4.5.2 Catalytic cracking

Cracking breaks large oil molecules into smaller ones. One application is in cracking a heavy gas oil to form high grade petrol and gas. The process is known as 'fluid catalytic cracking' – 'fluid' because the catalyst (this helps the cracking reaction without being changed itself) can be made to flow like a liquid when it is blown with air or hydrocarbon vapour.

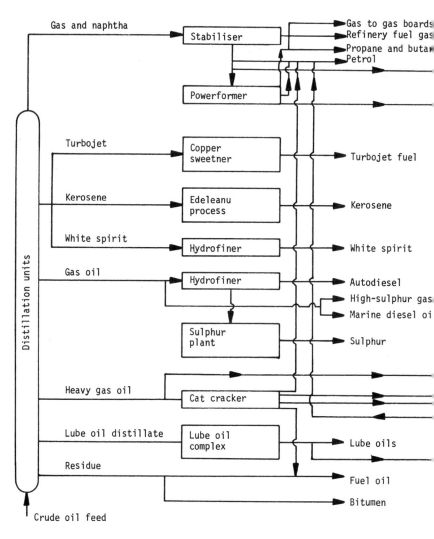

Figure 4.6 Simplified flow diagram of oil refining processes.

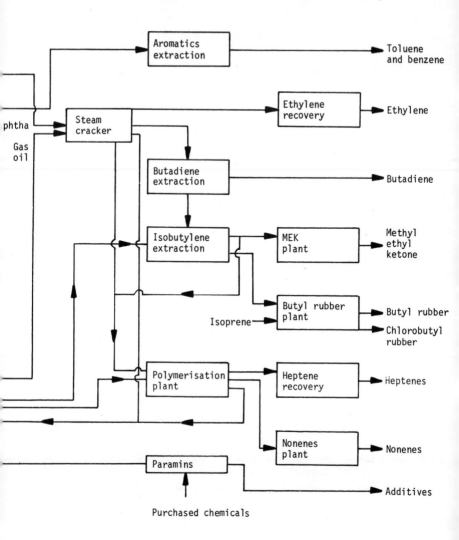

phtha

Gas
oil

Aromatics
extraction → Toluene
and benzene

Steam
cracker

Ethylene
recovery → Ethylene

Butadiene
extraction → Butadiene

Isobutylene
extraction

MEK
plant → Methyl
ethyl
ketone

Butyl rubber
plant → Butyl rubber
→ Chlorobutyl
rubber

Isoprene →

Polymerisation
plant

Heptene
recovery → Heptenes

Nonenes
plant → Nonenes

Paramins → Additives

Purchased chemicals

Courtesy: Esso Petroleum Company Limited

4.5.3   Powerforming

This process changes the configuration of atoms within molecules rather than the size of the molecule. The powerformer converts low-quality naphtha from primary distillation into a high-quality petrol component.

4.5.4   Polymerisation

Polymerisation is the reverse of cracking; it joins small molecules into chains of molecules. Light gases, produced by the catalytic cracker and chemical units, are combined to produce heavier materials such as heptenes and high-quality petrol.

4.5.5   Treatment processes

Almost all the fractions obtained by primary distillation need to be further treated to meet the stringent specifications demanded. One example is the removal of sulphur from such products as diesel oil and white spirit. This is done by a hydrofining process, in which sulphur compounds are converted to hydrogen sulphide. The H_2S is then fed to a sulphur recovery plant which produces pure rock sulphur. Further examples of treatment are the Edeleanu process for the removal of aromatics from kerosene and the copper sweetening process for converting corrosive and malodorous compounds into non-corrosive and inoffensive ones.

4.5.6   Lubricating oil manufacture

Feedstocks for lubricating oil manufacture are prepared by vacuum distillation of the residue from primary distillation of certain crude oils. There are three further processes:

1   Phenol treatment to remove undesirable aromatic compounds.

2 Hydrofining to improve the colour and the stability of the oil.

3 Dewaxing to improve the 'pour' characteristics.

4.5.7 Bitumens

On this unit the residue from vacuum distillation of Venezuelan crude oil is blended with a diluent to produce a range of penetration grade bitumens, some of which are subsequently air blown to produce oxidised bitumens.

4.5.8 Steam cracking and ethylene recovery

Steam cracking is a high-temperature thermal cracking process used for producing olefins and diolefins as feedstock for other chemical, plastic and rubber manufacturing processes. The feedstock of either naptha or gas oil is vaporised, mixed with steam, and heated to high temperatures. Under these conditions most of the high-molecular-weight hydrocarbons in the feed are cracked to produce a wide range of low-molecular-weight hydrocarbons including ethylene, propylene, butenes and butadiene. These are separated as saleable products or for further processing in other plants.

4.5.9 Isobutylene extraction and recovery

This extraction process, using sulphuric acid as a solvent, separates and recovers isobutylene from mixed butenes produced during the catalytic and steam cracking processes. Isobutylene is one of the feedstocks for the manufacture of butyl rubber.

4.5.10 Butadiene extraction

These units recover and purify butadiene from the raw

butadiene-containing streams produced by the steam cracking units. Butadiene is fed to a neighbouring plant for the
manufacture of the synthetic rubber — styrene-butadiene
— used for car tyres.

4.5.11   Butyl rubber unit

This unit consists of a copolymerisation section where raw
rubber is made from purified feed streams of isobutylene
and isoprene, and a finishing section, where the raw rubber
is dried, baled and packaged. Chlorobutyl rubber is also
made here.

4.5.12   Aromatics extraction unit

Another extraction process, uses a feedstock, rich in
aromatics from the powerformer. This feedstock is subjected to sulfolane extraction and subsequent fractionation
to recover toluene and benzene. These important intermediate chemicals are used in making plastics, synthetic
fibres, detergents, paints, dyes, etc.

4.5.13   Methyl ethyl ketone unit

Butenes from steam cracking are extracted with sulphuric
acid to recover n-butenes as secondary butanol (butyl
alcohol). The crude alcohol is purified by fractionation
and dehydrogenated, by a catalytic process, to methyl
ethyl ketone (MEK) which is then purified by further
fractionation. MEK is used as a solvent in industrial and
pharmaceutical applications.

4.5.14   Paramins plant

This consists of a series of vessels in which various purchased
and indigenous chemicals are reacted or blended, in batch

processes, to give a wide range of additives, a number of which are used to improve the specific characteristics of lubricating, fuel and crude oils.

4.5.15 Gas

Gases are either derived from the crude oil itself or from cracking reactions. In recent times they have been processed to form a gas similar to that obtained from coal. Both propane and butane liquefy under moderate pressure. The former is used for metal cutting and welding, and for domestic purposes in cold climates. The latter is an alternative to coal gas.

4.6 Plastics

The majority of plastics have oil as the source of their raw material and it is therefore appropriate to refer to these at this juncture.

There has been a growing concern that waste plastics may become a global 'pollution hazard', the term being used in its somewhat extended context.

As is so often the case of manufactured goods, plastic waste falls into two main categories: the first is unavoidable production waste, the second, by far the more troublesome, arises from discarded items, usually used once only which becomes mixed with general refuse during the ordinary course of events. It is this class of waste that the environmental conservationist feels should be recycled. The simple answer to this might be that the many grades of material encountered militate against recycling; there are other problems as well. If such waste is recovered within the same factory as it arises, the composition of the material is known and can be kept separate for recycling.

It is of value to refer to a recent report of The British Plastics Federation which considers the recycling, recovery reclamation and savlaging of plastic waste [4.6]. The report points out that the re-use of·plastics must take into account

the two distinct classes that exist; the thermoplastics and
the thermosetting plastics. Recycling of the former is currently
undertaken on a considerable, by and large viable, scale. This
material is much used in the packaging field, the containers
being printed with advertising matter. The inks used for this
purpose are difficult to remove before reclamation can be
carried out. A further trouble lies in the fact that, as now
collected, the containers are made of mixtures of dissimilar
materials which, although dissimilar in the property sense,
may be closely matched in terms of density and melting or
softening point, so that separation methods such as are used
in the metal field cannot be applied. Reclamation of thermoplastics is more likely to become established for the recycling
of scrap mouldings made from basic polymers where both
the material type and any additive content is a known
quantity. Additives for plastics is becoming as important a
field as the polymers themselves, since they effectively
stabilise the material under varying environmental conditions.

The thermosetting materials have a restricted reclaimability
for the reasons given above, so that they are limited to
grinding for use as a reinforcing filler.

One of the conclusions given in the BPF report states that
'Recycling can take place only when the material to be reused
is equal in quality and no greater in cost than virgin raw
material. It is these two constraints which effectively limit
recycling at present because of the cost of reclaiming or
salvaging followed by further cost of sorting, grading and
cleansing.'

4.7   Ferrous metal

Concisely, the different forms of iron and steel are all derived
from iron ores, which consist of iron in association with
various other elements, principally oxygen and carbon, in the
form of oxides and carbonates. The coke-fired blast furnace,
already mentioned, is used to smelt the ore to extract the pure

iron. A flux is added which combines with the impurities, forming slag, and separates most of the iron. This slag has valuable uses and is placed on the market in the following grades.

1 *Air-cooled slag* Much used as a concrete aggregate and as a medium for filter beds such as are used in sewage treatment. The material may also be coated with tar to form tar-macadam and asphalt.

2 *Granulated slag* This is used in the production of Portland blast furnace cement. It is also used in the production of certain classes of abrasive powders and in the manufacture of building blocks.

There is one other product that can be produced from the steel slag and that is a phosphate fertiliser. This is obtained in works where the phosphorous content of the iron ore is extremely high.

4.7.1 Cast iron

Cast iron is the product of the blast furnace process, and there are three grades, *viz.* grey, white and mottled, the difference being in carbon content. When cast iron scrap is used as one of the components of a metal mixture which is melted for the production of cast iron it is important to have knowledge of its analysis. It is advantageous therefore to be able to make a reliable assessment of the composition of cast iron scrap generally available. To assist in this matter, cast iron scrap can be classified into identifiable types of scrap.

Scrap used in foundries is normally obtained from recognised scrap metal merchants and it is they who grade the material prior to redistribution. For further information of the types of cast iron scrap and their classification, it is suggested the Institute of British Foundrymen's report [4.7] is studied.

4.7.2   Steel

There are a great many varieties of steel, the properties of which vary according to the percentage of carbon and of metals other than iron present, and also according to the method of preparation.

The problem for the processor of scrap is in dealing with these varieties, and this requires good knowledge of the subject. The type of scrap used in the open hearth and Bessemer processes varies with specification and a clause is often found in contracts excluding coated metals. For electric arc foundries, the exclusions may be phosphorus, sulphur or silicon up to a predescribed fraction.

Sources of scrap steel are mainly discarded plant, buildings and marine structures, although large quantities of miscellaneous steels are involved.

The use of cast iron scrap in steelmaking is relatively small and occurs mainly with small items. A small tonnage of cast iron borings is also consumed. Other lesser users are the blast furnaces — mainly in the form of borings mixed with steel turnings, burnt cast iron scrap and firebars.

McDonald [4.8] has reviewed the techniques and problems of using alloy scrap for steelmaking.

4.7.3   Wrought iron

This is the purest commercial form of iron. Much of the work of the scrapyard is spent in testing for wrought iron among the steel and, with a few exceptions, this can only be done by fracturing a sample for visual inspection.

4.8   Non-ferrous metals

The non-ferrous metals are used in industry principally in the form of alloys. There are a very large number of proprietary alloys such as Delta metals, phosphor and manganese bronzes, and various brasses and white metals which are produced by various firms and which are aimed at particular applications. As in the case of steels, the recovery and re-cycling of non-ferrous metals is complicated. The major problem, therefore, rests on segregation. A great deal of hand sorting is still relied upon but sophisticated chemical and electronic testing are carried out. Because of the high value of many non-ferrous metals and alloys the processors have the further problem of security whilst the metals are being sorted, processed and delivered to consuming outlets.

4.8.1 Aluminium

One of the best guides to the aluminium industry is by Pearson [4.9], although this is now about twenty years old. The industry operates processes introduced by Hall and Héroult about eighty years ago for the electrolytic production of aluminium and there have been few changes in the basic method in this period.

The ore principally used is bauxite and the alumina is extracted from it by the Bayer process. This produces a pure alumina which is dissolved in molten cryolite used as the electrolyte in the Hall-Héroult cell.

There are, strictly speaking, no by-products from the aluminium industry other than the main products and some waste products. The red mud produced as waste from the Bayer process has had some use as a red pigment. The whole subject of utilisation of red mud residues from alumina production was researched by the USA Bureau of Mines, the findings of which were published in 1970 [4.10].

Alumina is used for refractories, abrasives and as a starting point for aluminium chemicals and most major aluminium companies have a chemical division marketing these products. Whilst the move into this area results from their own expertise and knowledge of the Bayer process it is normal for the aluminium hydrate and alumina to be produced with different operating conditions in order to yield a product suitable for these end uses. Smelter grade alumina is usually a larger scale product cheaply produced under conditions applicable only to itself.

Hydrated alumina may be sold for alum manufacture, and dross from metal melting is sold to enable processors to separate aluminium and aluminium oxide components.

The recycling of scrap aluminium and its alloys presents similar problems to those encountered in re-using other scrap metals. The problem is to collect and treat economically these varied materials and to produce from them alloys of practical value and standards of quality. It would seem that the most attractive method would be to separate the aluminium alloying fractions and to use the pure and lightly alloyed aluminium so obtained to make whichever alloys are required. This is, of course, possible, but unhappily the cost of the alloys so produced would likely to exceed the cost of those made from primary aluminium. Moreover, it is economically desirable that the alloying fractions present in the raw material to be refined should be again used in producing new alloys. This would raise the question of separation of the individual metals from the residue and from most of the known processes this would likely inflate cost and involvements. Smith [4.11] has discussed these and other related matters.

4.8.2 Copper

The production of by-products from the smelting and refining of primary copper depends on the type of ore being processed and thus varies considerably throughout the world. In general, however, they consist of other metals such as cobalt, tellurium and selenium and precious metals in the form of gold and silver. Nickel may also, in some instances, be considered a by-product of copper, though in certain areas of Canada, for instance, both nickel and copper are extracted as primary metals from the same ore. The Sudbury mines in Canada annually produce some 160,000 tons of nickel, 135,000 tons of copper and over 160,000 tons of high-grade iron ore with a by-product yield of over a million ounces of gold and platinum and 1.5 million ounces of silver.

In many copper smelting plants sulphuric acid is also produced from the sulphur dioxide gas removed from the matte during conversion. Some plants, however, use the sulphuric acid obtained for their own leaching processes. As a result of the world-wide concern for pollution and development of 'clean' converters and flash smelting (in the USA and Finland, for example), the production of sulphuric acid as a by-product of copper may well be considerably increased.

The precious metal by-products are mostly obtained from the slimes removed from the electrolytic refining tanks.

To some extent copper is also a by-product of its own processing in that in some instances it is possible to extract further metal from the tailings or waste carried away from floatation cells. This is generally carried out by leaching.

Some copper may also be recovered from smelting furnace slag. The slag is sometimes used for road making or as an abrasive material for sand blasting.

The recovery of scrap, or secondary copper, is a very important aspect of the whole copper consumption contributing to some 40 per cent of Britain's annual consumption of the metal. There are several firms who specialise in the recovery of secondary copper and whose equipment — particularly that used in the final refining operations — embodies the very latest processing and techniques. Much of the scrap copper and copper-based metals collected in this country are re-used with the minimum possible amount of refining. Refining operations seldom result in complete separation of the constituents and in the process of removing the impurities, significant quantities of copper are expected to be lost. It is for this reason that refining operations are kept to a minimum. Mantle and Jackson, who have discussed copper scrap reclamation in detail [4.11], point out that alloys have to be produced to certain specifications with limits set for impurities, and the attainment of these specifications, if expensive operations are to be avoided, is only possible by good scrap selection. The emphasis is, therefore, on good sorting by merchants.

4.8.3 Mercury

The recycling of used mercury is a regular operation of the industry concerned with refining this expensive metal. The used metal is reclaimed from out-of-date scientific apparatus, from discarded electrical appliances and from processes making use of the metal as a production aid or catalyst. The metal is collected by scrap merchants and sold to the refiners or it is returned by the users themselves for retreatment. It is estimated that about 5 per cent of the mercury on the market comes from this source. Used mercury is sometimes returned to the refiners with a valuable contaminent such as silver in significant quantities. Such contaminents are salvaged when it is economically justifiable to do so.

4.8.4 Tin

This is a soft, malleable and ductile, silvery-white metal. It is extracted by heating tin oxide with powdered carbon in a reverberatory furnace.

There appear to be no major by-products arising from the extraction and refining of the metal, although where mining is carried out by tunnelling through, say, granite (as in Cornwall), the stone is sold off as chippings. This however, does not apply to the larger producers where the tin is alluvial.

The most important single use for tin is in the manufacture of tinplate, and the bulk of the metal used for this application is not reclaimed. Reclamation is, however, economic from processed scrap during the manufacture of cans. Some secondary tin arises from white metal bearings, solders, etc., and this is normally re-used in an alloyed form rather than refining it to pure tin metal. Again, however, much of the tin is not reclaimed, since usage in an application such as soldering disperses it widely. A high percentage of the tin is probably re-used when it is in the form of bronze, since many of the lower grade bronze alloys are made with secondary metal.

4.8.5 Nickel

Nickel is a silvery-white metal used mainly as an alloy addition to other metals and for plating. It occurs in ores found in many parts of the world. Mining and extraction methods depend on the mechanical, physical and chemical characteristics of the particular ore, and these have been discussed by Boldt and Queneau [4.13]. Some ores contain nickel and copper together. Other valuable elements sometimes present are the platinum group of metals — gold, silver, cobalt, selenium, tellurium, iridium and sulphur — which are recovered. Platinum is used for electrical contacts, scientific instruments and as a catalyst. Palladium is used in high-temperature brazing alloys. Rhodium is used as a hardener for platinum and for plating purposes. Iridium is extremely hard and resists attack by many other metals.

4.8.6 Lead

Briefly, there are but few by-products in the lead manufacturing industries. When lead is melted a small amount of dross is formed and this is collected (for health reasons) and sold to smelters. Such activities arise in the manufacture of electric cable sheathings, casting of electric battery components, shot, sheet, piping, foil and castings production. Other industries recovering lead dross from molten lead baths are those concerned with coating steel wire with lead and the production of tetraethyl lead (metallic lead arises as a by-product which is remelted within the factory and some dross results from this).

Lead blast furnaces, used for reclaiming waste lead products, produce a slag containing a low percentage of lead, and a lead refinery produces drosses which are usually processed to recover other metals.

The sources of scrap lead are: (a) lead pipes, sheeting and fittings; (b) battery lead which has been used in the automotive industry; (c) rejected lead, which is often in finished form; (d) sheathing from discarded cables. The compositions of these materials are such that some segregation is required:

4.8.7   Zinc

This soft bluish-white metal is extracted by roasting zinc-bearing ore to form the oxide, which is then either reduced with carbon and the resulting zinc distilled, or reduced electrolytically. The metal is extensively used in alloys, especially brass, and for galvanising iron.

The by-products arising from zinc manufacture are sulphuric acid, cadmium and sometimes indium. Silver and gold are also present in zinc/lead minerals but are mostly recovered with the lead. Some cobalt is also produced by certain electrolytic zinc refineries. The winning of these by-products is well covered in metallurgical text books.

Sulphuric acid is a valuable by-product of the roasting process. Some is often used at the zinc works to make ammonium sulphate and calcium superphosphate fertilisers. In the latter case, gypsum is a further by-product used to make plaster for building construction. Sulphuric acid is also an important raw material for many other chemical processes.

The earliest form of zinc dust was known as 'blue powder' and this originated as a by-product of the horizontal retort process for the production of zinc. Most zinc dust today is produced by distillation. Zinc or high grade scrap, such as galvanisers' dross, is evaporated in a closed system and condensed to form a fine powder. The conditions of evaporation and condensation can be controlled to give consistently a dust of the required particle size. Manufacturers can usually offer several grades of dust of differing particle size characteristics and chemical purity. Zinc dust is used as a pigment in paints for the protection of iron and steel against corrosion and as a reducing agent for a variety of processes in chemical industries.

4.8.8   Aluminium oxide and silicon carbide

Both of these materials are the concern of the abrasive trade. The first is a crystalline substance and the latter a compound. Silicon carbide is known by the trade name 'carborundum'.

Both materials are formed in electric furnaces.

In processing aluminium oxide, and with special reference to grinding wheels and allied items, the chief waste arises from the finishing operations. The waste is mainly composed of abrasive grain in various grit sizes mixed with the bonding agent. These grains find uses in cement mixtures for moulding and forming non-slip surfaces. Finished production scrap can be broken down and the bonding agent (in certain cases) burnt off to form a refractory material. A large proportion of abrasive articles are bonded by a ceramic process and if it is desired to retrieve the material it can be done by crushing and sifting techniques. But the grain remains with the bonding on and this restricts its use as a high-temperature refractory, although it can be so used in the lower temperature ranges. This waste can also be used for making non-slip pavings and as a pressure blasting grain.

There is probably more waste incurred by the actual users of aluminium oxide articles than in the manufacture of them. A fairly large percentage of grinding wheels and segments made from this material are scrapped after they have worn away to an unusable size. This waste could well amount to twenty per cent of the material in the first place. Some users of grinding appliances granulate the cast-offs and use them in tumbler barrels for fettling and polishing castings and forgings.

Silicon carbide fines can be used as an additive to silicon carbide refractories which are designed to withstand high temperature. Used refractory scrap — usually emanating from kiln furniture — can be crushed and used for making a non-slip surface on concrete floors.

4.9 Asbestos

This is a fibrous variety of hornblende, separable into flexible filaments and much used in constructional work and for fireproofing.

The industry as a whole has over many years developed to a considerable extent the re-use of so-called left-overs from the various processes. The nature of the material has lent itself well to this since the very wide range of products provides the opportunity for using asbestos fibres in an equally wide range of fibre lengths. Commencing with long fibres for products such as asbestos textiles where a long staple is necessary, short fibres which become by-products from this process are suitable as the prime material for other products and processes where length is of no importance, and so down the scale to those products such as floor tiles where the shortest of mined fibres are suitable. In general, therefore, it is fair to say that the re-utilisation of waste material has already been developed to a considerable extent. The material which has to be disposed of has, also, reached a point where the residual value is very low in relation to the economics of reclamation.

4.10   Carbon and graphite

Graphite is a crystallised form of carbon which occurs as a mineral in various parts of the world. There is also a synthetic variety, much used in industry, that is manufactured by baking petroleum coke in a controlled furnace. With this we are more concerned.

The manufacture of carbon and graphite gives rise to some by-products, both from the product itself and from the coke packing medium used in the process. The basic material used in both of these areas constitutes a high price element and therefore much work has gone into the reclamation of waste. Waste capable of re-use is returned to the process for further use. For scrap not so absorbed a number of markets have been developed for its sale. This applies to milled graphite where off-cuts or turnings from machining are furnace electrodes and special blocks are used to produce grades of milled powder. Other by-products include material which arises from the screening of coke packing medium in the form of granular, fines or dust with a suitable carbonaceous content.

All of this waste, whether as powder, lumpy graphite scrap or screened pack material, is marketed in various ways:

1   Carbon content items for the iron and steel industry,

e.g. recarburisers.

2 As a raw material for other carbon manufacturers, e.g. battery carbons and welding rods.

3 As moulding dressings for the foundry industry.

4 As fillers for the plastics industry.

5 For lubrication (graphite powders) — generally specialised lubricants.

6 As anode additives for the aluminium industry.

7 Carbon black used in the rubber industry, particularly for tyres.

Some waste from specialist manufacturers of carbon products is mixed with copper in various percentages. This material is sold to copper refiners who extract the copper content.

4.11 Chalk

This well-known calcareous earth or carbonate of lime is mostly used for the manufacture of cement.

The by-product from digging out the material is flint. Experiments with chalk flints were being made as far back as 1914 for the manufacture of tiles, but it was two years later that they were used in general earthenware. Because of modern techniques in chalk digging the production of suitable nuggets for this purpose has declined and are now available mainly from lime works where hand-sorting is still carried on.

Washmill flints were introduced about 1926 and their use increased gradually up to 1939. With the outbreak of the war the cement companies became practically the sole source of supply to potteries.

The sections of the pottery industry which are the main users of flints are: white glazed wall tiles, tableware and sanitary ware.

4.12   Fuller's earth

This loosely applied term is more correctly reserved for absorbent clays composed almost entirely of the natural clay mineral, calcium montmorillonite. No definite chemical composition can be given to the material, or even its principal mineral, montmorillonite. Its usefulness is due to its high surface area which, weight for weight, is far greater than that of other minerals. If 'activated' by treatment with a mineral acid, this area can be multiplied to something over three hundred square metres per gramme.

Its traditional employment was with the textile industry, where it was used for 'fulling', and to its other age-old application in pharmacy. Modern uses are in the refining of edible fats and oils from which the earth removes colour and impurities, thus imparting a share to the appetising appearance for these fats and oils. It is also employed in the refining of technical oils and waxes, mineral oil refining and also in the regeneration of used lubricating oils.

When converted to the 'bentonite' form, it is used in civil engineering and oil-well drilling and for many years has been used in iron foundries as a bonding clay in sand moulds.

Certain waste by-products occur from the 'activation' treatment, these consisting either of (a) wet cake of calcium sulphate with iron and aluminium hydroxides or (b) iron and aluminium oxide. No useful by-products occur in the preparation of the earth for any other purpose.

4.13   Glass

This hard, brittle, transparent or transluent substance is a mixture of the silicates of calcium and sodium obtained by melting together silica, calcium carbonate and sodium carbonate. Special-purpose glass may contain lead, potassium, barium, aluminium and other metals in place of part or all of the sodium.

No by-products, as such, arise during manufacture, but cullet (scrap glass), more especially if the composition is known, and production waste is re-melted for further use.

Among the lesser known uses for glass is in the manufacture of glass wool, a product used for filtering and
absorbing corrosive liquids, as well as an insulator for
sound and heat.

The re-use of cullet is an important aspect of the glass
industry, especially during the prevailing conditions. For
advantage in re-processing, manufacturers prefer to use
cullet derived from their own products, but this is not always practical. Cullet is collected from miscellaneous
sources and a mixture of classes of glass ensues. Nevertheless,
vast quantities of cullet are diluted with raw material to
give an end product highly suitable for the majority of
purposes.

There is also a developing demand for powdered glass —
a material processed by glass millers. Their raw material is
mixed cullet furnished by local authority cleansing departments and by merchants. It is cleansed, classified into
colours, and then ground or screened into particles sizes.
Outlets for the finished products are:

Abrasive cloths and papers	—	Amber
Abrasive products	—	Amber and white
Match heads	—	Amber and white
Matchboxes	—	Amber
Glazed fireclay products	—	Amber
Gear cutting	—	Amber and white
Foundry fluxes	—	Amber and white
Decorative wall finishings	—	All colours
Flooring systems	—	All colours
Steel extrusion	—	White
Paints and compositions	—	White
Moulded sculptures	—	All colours

Blue glass, originating from bottles used for containing
poisons, is the main source of blue cullet, and when
coarsely ground forms an attractive reflective medium for
fascia tiles. Amber and green glass, sometimes mixed, are also
used for this purpose.

One of the impressive uses of scrap window glass, commenced in 1941 by Jacques de Sejournet, is as a lubricant and

heat insulator in the hot extrusion of steel. His work has led
to advances in hot-metal-forming techniques.

Collectively glass manufacturers have a forward-looking
programme for the future re-use of an extra 100,000 tonnes
of cullet per year. This will mean that almost 30 per cent of
every bottle will be made from re-cycled material. Whilst
recycling is one of the key recommendations of a Glass
Manufacturers' Federation report [4.14], the industry is
also to sponsor research of new uses for crushed glass. A
programme is to be financed at the University of Cardiff's
Wolfson Laboratories aimed at developing markets for
cullet outside the glass container industry. Such uses as
road surfaces, flooring materials and wall construction are
envisaged. The initial industry target can be met from
existing sources, which include used and damaged bottles
from bottlers, dairies and cullet merchants. Larger supplies,
however, will mean new sources. The answer is likely to
be in the development of a new central and local government
policy with regard to solid waste disposal.

4.14 Granite

This is a heterogeneous mixture of felspar, quartz and mica,
and its quarrying and dressing for commercial markets gives
rise to large volumes of waste in various sizes and shapes.
The rubble is graded and marketed as granite chippings with
outlets to the civil engineering and building trades. Veneer
waste is sold in random sizes for the construction of fire-
places, crazy pavings and ornamental items. Waste of roughly
cubic shape is disposed of as edgings to paths or walling
stone. Other random material is shaped into gateposts,
lintels, cills, table tops and ledges. Dust arising from working
granite is widely used as a jointing element to give a uniform
appearance between granite blocks in constructional work.
Nevertheless, despite the considerable trade that has been
built up to profitably dissipate waste, the percentage of loss
is never low enough.

4.15 Kaolin (China Clay)

Kaolin is a white clay derived from the decomposition of the felspar in granite. This important versitile material has many desirable qualities, including softness, inertness, opacity and other factors. It is mainly used as a chinaware ingredient and in the manufacture of paper. It is also used in a wide variety of industries, such as paint, rubber and plastics, pharmaceutical and cosmetics, insecticides and fertilisers, inks and many others.

As stated, kaolin is the product of the alteration of felspar in the granite, but in some areas only a certain degree of kaolinisation has taken place. In some areas none has occurred at all. This, therefore, provides the basic minerals — kaolin, quartz, felspar and mica. Other minerals occurring either in smaller proportions or in localised regions, include tourmaline, topaz, fluorspar, cassiterite, etc., but for the most part the economic interest in these is small.

Kaolin being the main product will not be enlarged upon as our interest lies in the by-products field. Therefore the remaining materials which are at present exploited are:

1   Felspar of felspar/quartz (China stone).

2   Quartz (sand).

3   Mica.

China stone as quarried is a mixture of felspar, quartz, mica and small quantities topaz/fluorspar. This is used to produce a material called DF stone which is a mixture of felspar and quartz, the topaz, fluorspar and mica having been removed by a floatation process. The removal of these impurities is to meet a number of technical and environmental objections.

DF stone is used in the manufacture of grades of earthen ware, vitreous sanitary ware, porcelains, wall tiles and bone china bodies.

The sand is the main by-product of the kaolin industry and is used in the building and civil engineering fields. Most grades of building sands and gravels are produced, from the fine plastering sands to large size concrete gravels. Some processors of China clay produce their own concrete products from these waste sands, including building blocks, panels, cills, mullions beams and so on. Calcium silicate bricks are also made from a mixture of high quality silica sand, silica flour and burnt lime.

Micas differ in chemical composition, but they agree in their character, all splitting into thin, flexible plates or laminae. The chief varieties are muscovite (potash mica) and biotite (magnesia mica). The heat-resisting qualities of micas enables the plates to us used as glass for lanterns and heating appliance doors. In the electrical field they are used for iron and toaster elements. Grades of mica powder are used in the wallpaper trade to give a silvery sheen. Powders are also applied in the manufacture of paints and certain rubber goods, and to assist the adhesion of textured wall and ceiling surfaces. These powders also find a use in oil well drilling muds, welding electrodes, electricity cables, sound-proofing compounds, roofing felts and flame-proof insulators.

4.16   Marble

The compact and mottled varieties of limestone are commonly known as marble. The industry's main and traditional products are blocks and slabs for building construction, monuments, sculpture and decorative items, both domestic and industrial. Preparation of the material for the market gives rise to considerable production waste, much of which may be disposed of in granulated form or as a powder.

Crushed and graded marble is used for an aggregate for terrazzo and exposed panelling, for the reconstruction of marble products and, to a limited extent, for the production of hydrogen. Production waste is also used as a substitute for less readily available and cheaper materials, e.g. concrete aggregate, roadstone and cement manufacture. But these uses are mainly limited to areas where marble is plentiful and the more usual materials are not so. Now and again a marble deposit will prove unsuitable for block production because of naturally occurring inferiorities, usually connected with colour. This defective material is used for various industrial purposes.

Dust is used as a filler for paints and plastics, in bio-chemical culture beds, in cosmetic products and as a base for fertilisers.

4.17 Slate

The several kinds of this argillaceous stone of laminated structure are mostly employed in building construction, but there are other uses, especially where corrosive substances are encountered. The material is shaped to form various architectural details and, in other industries, slabs and special forms. All of this gives rise to enormous waste, sometimes well over 60 per cent of the whole, although substantial reductions can be made by the sale of walling stone from material which would otherwise be tipped. Other waste arises from sawing and finishing the items referred to, as well as from the dressing of roofing slates, even though a large range of random size slates are made.

Sawdust and off-cuts are used to produce slate granules and powders. Lightweight aggregate is produced by heating slate waste in rotary kilns and crushing. In this form it is suitable for filter media in sewage works and the like. A much-used by-product is slate powder used in bituminous mixes for asphaltic flooring, bituminous paints, damp-proof courses in buildings, mastics and roofing felts. It is also used in the manufacture of insecticides, pigment powders, plastics, fertilisers and sealing compounds.

4.18 Sulphur

The two most important sources of sulphur for industry are:

1 Frasch sulphur.
2 Sulphur recovered from hydrocarbon processing.

Volcanic deposits of native sulphur, whilst once important, are now of little commercial significance. Sulphide ores, with the exception of pyrites, are exploited primarily for the value of the metals they contain although it is often convenient to simultaneously use the sulphur content in the manufacture of sulphuric acid.

Frasch sulphur is the type of native sulphur first-exploited by the Frasch process in the USA. This sulphur occurs in 'domes' of mineral salts usually between limestone cap-rock and the underlying calcium sulphate layer, which in turn is usually situated over a dome of rock-salt.

With the growth of the world petrochemical industry and the need to exploit sources of oil and gas containing increasingly higher impurity contents (including sulphur compounds), sulphur removal and recovery has become an essential part of oil and gas processing. The methods of recovery of sulphur from oil and gas processing are essentially simple. Hydrotreatment of oil fractions produces hydrogen sulphide; hydrogen sulphide is also present in some natural gas. Hydrogen sulphide is recovered from gas streams in both cases, usually by amine washing. In this way fairly concentrated streams of H_2S are obtained which are then processed to elemental sulphur by the well established Claus principle. Sulphur production in Canada – a major market supplier – is almost exclusively from gas and, to a lesser extent, oil. Other important gas-based sulphur production industries exist in France (Lacq) and the USSR (Orenburg).

The major use of sulphur is in the manufacture of sulphuric acid, which in turn is chiefly used in the manufacture of phosphoric acid and phosphate fertilisers. Other important industries (tonnage-wise) include titanium dioxide, caprolactam detergents (again sulphuric acid uses), carbon disulphide manufacture and dyestuffs.

The by-product of phosphoric acid production is calcium sulphate. This has hitherto been rather a problem but the technology exists now to convert this back into sulphuric acid, make it into ammonium sulphate, or merely to produce plaster board and wall-blocks from it.

Ammonium sulphate is a by-product of caprolactam

manufacture. This material is a valuable fertiliser, particularly in view of the present tight market situation. Ammonium sulphate is also produced although in dwindling amounts from coke-oven gas (derived from the sulphur and nitrogen impurities of coal).

Production of titanium dioxide produces ferrous sulphate as a by-product. This, too, can be rather troublesome to deal with, although again it it possible to convert it into sulphuric acid, providing the right kind of roasting equipment is available.

The carbon disulphide industry produces hydrogen sulphide as a by-product which is normally converted back into elemental sulphur and re-cycled through the manufacturing process.

Sulphuric acid productions results in the evolution of sulphur dioxide to the atmosphere and is one of two chief sources of SO_2 emission, the other being electric power generation (from sulphur-contaminated fossil fuels). There are many ways of reducing the level of these emissions to very low figures. In the case of sulphuric acid manufacture these methods chiefly result in increased sulphur utilisation, whereas in the case of power generation such useful by-products are not normally recovered, with the result that the incentive to control SO_2 emission in the power industry is low. Legislation, however, is tending to provide the drive.

4.19 References

4.1 Sir Kingsley Dunham, 'Global resources of metals for the electrical industries', *Electronics & Power* (November 1972).

4.2 C. Lapworth, *An intermediate textbook of geology,* William Blackwell & Sons.

4.3 R.J. Dower, 'Reconstitution of waste metals,' *Chartered Mechanical Engineer* (June 1973).

4.4 *Coal preparation,* Seeley Mudd Series, American Institute of Mining Engineers.

4.5   *Ash and agriculture,* Central Electricity Generating Board.

4.6   *Recycling, re-use and recovery of plastics,* The British Federation.

4.7   *Report on cast scrap iron,* Institute of British Foundrymen.

4.8   R. McDonald, *The use of alloy scrap for steelmaking,* British Steel Corporation.

4.9   T.G. Pearson, *The chemical background of the aluminium industry,* The Royal Institute of Chemistry

4.10   Bureau of Mines Report of Investigations, United States Department of the Interior (November 1970).

4.11   F.H. Smith, *The recovery of aluminium from waste products,* Society of Chemical Industry.

4.12   E.C. Mantle and N.H. Jackson, 'The reclamation of scrap', *Copper Development Association Journal* (January 1968).

4.13   Bolt and Quenau, *The winning of nickel.*

4.14   *The glass container industry and the environmental debate,* Glass Manufacturers' Federation.

4.20   Firms, societies and other bodies consulted

Aluminium Federation

Amoco Europe Incorporated

Anglesey Aluminium Construction Limited

Arcorundum Grinding Wheel Company

Asbestosis Research Council

The British Aluminium Company Limited

The British Petroleum Company Limited

British Quarrying & Slag Federation Limited

British Rubber Manufacturers Association Limited

British Steel Corporation

British Sulphur Corporation Limited

British Tar Industry Association

Burlington Slate Limited

Carborundum Company Limited

Cement Marketing Company Limited

Chalk Products Limited

Copper Development Association

English China Clays Sales Company Limited

Esso Petroleum Company Limited

Harris Clark Limited

International Nickel Limited

Institute of Geological Science

Laporte Industries Limited

Lead Development Association

Micafine Limited

National Coal Board

J. & H. Patteson Limited

Royal Crown Derby Porcelain Company Limited

Yarsley Research Laboratories

Steetley Company Limited

P.H. Stott and Company Limited

Sutcliffe Speakman and Company Limited

Tin Research Institute

Thomas Group (Marble and Granite)

Thomas Hill-Jones Limited

Universal Milling Company Limited

Zinc Development Association

Chapter 5

WATER

It is hardly reckless to suggest that water supplies override all other considerations. Vast sums are devoted to its conservation, purification, distribution and disposal. It is most likely that the demands for this commonplace yet vital substance will continue into the foreseeable future. There are no synthetic equivalents to help out.

There is a definite emphasis on re-use in the Government's far-reaching new proposals for the future control of this commodity. It is recognised that waste water, when suitably treated, makes a valuable contribution to the common stock of water resources. The direct re-use of waste water has existed for ages, sometimes with unsatisfactory results. Properly treated reclaimed water is now used for a wide range of important ends such as industrial cooling, various industrial processes, irrigation for some kinds of crop and for recreational employments. Further, there is a likelihood of an increasing use of reclaimed water in schemes for artificially recharging water-bearing strata liable to depletion.

5.1 Industrial cooling

The *barbarous* 'once-through' system is all too frequently used in industrial cooling. This water, often potable, after circulating through a cooling system, is discharged to waste and the heat so exchanged uselessly dissipated. If the water is discharged to public sewers eventual repumping ensues. Millions of gallons of water are still wasted in this way, mostly because of simplicity and convenience.

The recirculation of cooling water can almost eliminate wastage of water. Once the system is charged there is no

need for more water other than for topping-up purposes.
The system involves the continuous drawing of water from
the cooling equipment or process, passing it through a cooling device and returning it to the equipment or process. There
are various ways of reducing the temperature of the used
cooling water, the simplest being cooling ponds or spray
ponds. More sophisticated equipments include natural and
mechanical draught towers, air blast coolers and water
chillers. They may be expensive to instal, but are most
compact and have higher overall efficiencies.

5.2   Combined heat and electricity generation

Large-scale electricity generation usually involves heating
water to produce steam at as high a pressure and temperature
as possible and making this do work in a turbine. Afterwards
it is exhausted at as low a pressure and temperature as is
practical. The exhaust steam is then condensed and re-used
for boiler-feed water. A considerable amount of cold water is
required by the condensing plant to achieve this.
     The Central Electricity Generating Board draw their
cooling water from rivers, estuaries and coastal waters, to
which it is returned (at a higher temperature by reason of
its use in condensers) by pumping. Where rivers, and
especially canals, are used and the supply of water is limited,
or where temperatures may give rise to atmospheric or
thermal pollution, it becomes necessary to use cooling towers
for recirculating purposes. Whatever action is taken the
net result is an enormous wastage of heat.
     Although long technical struggles have gained minor
improvements in the efficiency of turbo-generator plant,
there are to be had large gains in the overall thermal
efficiencies of the plant by utilising the waste heat they
create. Some progress has been made in this field, notably
at Battersea Power Station, where waste heat is used to
meet the needs of the Pimlico district heating network. This
project has been described by Diamant [5.1].
     Quite obviously any waste heat must be flexible to meet
seasonal changes and it is highly desirable that the place of

usage should be adjacent to the power station. Many of our electricity generating stations are situated in the country and although there are perfectly good reasons for this it does mean that there is often no local outlet for waste heat. On the Continent the position is much more favourable, and many power stations are near enough to developed areas for hot water to be piped beneath the streets for heating purposes. There are no technical difficulties that cannot be surmounted, but it is hardly ever done in this country because of the absence of provision for it in the Terms of Reference of the CEGB. Nevertheless, it may well be that some factories — those concerned with heating processes — will be established in the country in order to take advantage of the surplus heat available.

5.3 Fresh water from salt water

Another use for the supply of waste heat from power is in the operation of sea-water distillation plants. Under consideration are design studies for such plant to be run on exhaust steam from turbines of an advanced gas-cooled reactor nuclear power station. Such a combined dual-purpose plant would therefore produce on a large scale electricity and fresh water.

The primary purpose of a sea-water distillation plant is, of course, to produce fresh water. However there could be a case for recovery of salt from the brine blow-down, but in most instances it is not commercially viable. In the Arabian Gulf State of Kuwait, where there is a large concentration of multi-stage flash sea-water distillation plants, it is a fact that the brine blow-down from the distillation plants is used as raw feed stock to salt evaporators and the salt so produced appears to be of good quality. It is not known to the writer if it is really an economic proposition since the idea was originally put into practice to prove the feasibility of salt factories using sea-water as the source.

5.4 Ground-water recharge

This relates to the recharging of water to the ground other

than by natural infiltration. The object is to increase the amount of ground-water held in storage and to allow an increase in rate of abstraction to take place. The method of replenishing may be by distributing water over the surface above a water-bearing bed or by injection of water through ditches, pits and wells. Recharge schemes allow for the disposal of storm water and pre-treated trade and other effluents.

Section 4 of the *Water Resources Act, 1963*, states that it shall be the duty of each river authority to ' . . . take all such action as they may from time to time consider necessary or expedient, or as they may be directed to take by virtue of this Act, for the purpose of conserving, re-distributing, or otherwise augmenting water resources in their area, of securing the proper use of water resources in their area, or of transferring any such resources to the area of another river authority.'

This and other sections of the Act seem to suggest that artificial recharging programmes are likely to receive more attention than hitherto. A paper by Buchan [5.2] and another by Taylor [5.3] throw considerable light on this controversial subject.

5.5   Process effluent

The problem of treating raw effluent from processing plant is often an individual one making it difficult to generalise. It is, however, possible to roughly divide the processing into stages. These usually comprise pre-treatment, primary, secondary and tertiary treatment, and sludge disposal. To obtain an acceptable standard any one or a combination of these may be necessary. Figure 5.1 shows the stages and sequence in a typical effluent plant.

Primary treatment may involve pre-screening and grit removal, blending and storage of effluent from various sources and the separation of oily or fatty substances. These steps may require to be followed up by chemical treatment, neutralisation and chemical addition to promote coagulation, and thence flotation, sedimentation or filtration to remove excess solids.

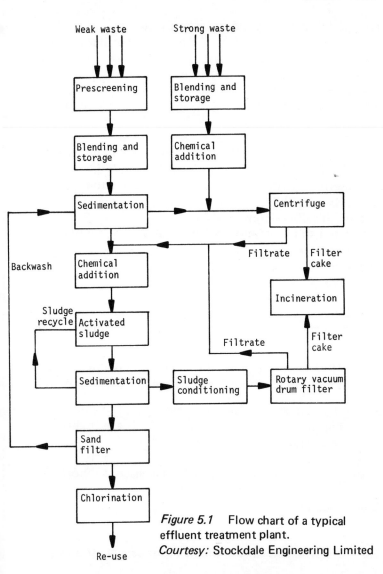

Figure 5.1 Flow chart of a typical effluent treatment plant.
Courtesy: Stockdale Engineering Limited

Secondary biological treatment involves completely mixed activated sludge processes, extended aeration, aerobic and aerated lagoons, filters and water stabilisation pools.

Tertiary treatment is employed when a high-grade effluent is desired for disposal or re-use. This may involve carbon absorption, stripping and other refinements.

For reasons already given the reception of sludges into waters and underground caverns most likely to cease, if it has not already done so. The trend, therefore, is towards sludge treatment to reduce moisture and, in the case of pressing, volume. Air-drying in lagoons is one way of de-watering, the solids being periodically dug out. De-watering by centrifuge is mainly used where filters are not suitable. Rotary vacuum drum filters are continuous in operation and are suitable where large volumes of solids are concerned. De-watering by filter press is more applicable to cases where relatively small batch volumes of sludge are involved, or where the solid content is small.

Filtered solids are usually tipped at authorised places, or, where there is any useful calorific value, incinerated. Filter cake is in a form that permits valuable materials to be recovered. A few examples are as follows:

1   Filter cakes from neutralised pickle liquors, etc., are sometimes processed in order to recover nickel, chrome, copper, etc.

2   Effluent from high-quality paper-making processes can sometimes be re-used for low grade paper or board.

3   Filter cakes from food processing applications and from pharmaceutical manufacturing plants are frequently used as ingredients in cattle food.

In two papers Barber [5.4, 5.5] has discussed both effluent treatment in processing industries and special filtration techniques.

5.6   Electrochemical treatment of industrial effluents

A variety of electrochemical treatments are now available for rendering toxic effluents harmless. Some of these processes can be combined with recovery of valuable metals, thereby making them more attractive to the industrialist.

Effluents containing metals in solution frequently occur

in industry, and they arise from various sources: plating, pickling, photographic processing, etc. Since electrolysis is the conventional purification process of many metals, it follows that this should be examined as a means of treating metal-laden effluent solutions. Electrolysis has been employed for some time to reduce the loss of precious metals, particularly silver, which has been recovered from photographer's 'hypo' (sodium thiosulphate). The recovery of silver is, of course, lucrative, but the recovery of tin, gold, copper, nickel, cobalt, cadmium can also be profitable.

There are a number of techniques on the margin of electro-chemistry which have proved effective in concentrating effluent solutions, electroflotation being one. This process, it is claimed, offers an economic solution to a wide range of industrial effluent problems, and is now being marketed by some of the major effluent plant manufacturers. It is a process that can be used in the treatment of emulsion paints, latex inks, millboard, etc., and for dealing with animal by-products and wool scouring. Basically this technique involves electrolysis of the effluent itself to release bubbles of hydrogen and oxygen which carry suspended matter to the surface, forming a sludge which is scraped off by a scraper conveyor. The process can also be used to de-water sludges, considerably reducing the volume, so making disposal easier and cheaper.

Surfleet [5.6] has discussed the electrochemical approach to the treatment of a variety of industrial effluents.

5.7   Water softening

Water softening in bulk may be achieved in a number of ways, and of these a few water authorities still employ the lime process. Water derived from chalk formations is not expected to lather immediately and hence softening takes place. Very briefly, the lime process involves high quality lump lime being added to an holding tank containing purified source water to form lime-water. This solution is mixed with the source water as it is pumped. This results in a chemical reaction whereby the calcium bicarbonate hardness is precipitated as calcium carbonate in the form of a thick sludge. The sludge is then filter pressed to a cake, dried and screened to form a powder. The result is commercially known as precipitated chalk and is mainly used in pharmaceutical and food outlets and as a filler.

5.8   References

5.1   R.M.E. Diamant, *Methods of centralised heat generation,* The Institution of Heating and Ventilating Engineers.

5.2   A. Buchan, 'Artificial recharge as a source of water', in *The problem of ground-water recharge,* The Institution of Water Engineers and the Society for Water Treatment and Examination (1964).

5.3   L.E. Taylor, 'Special reference to the London Basin', in *The problem of ground-water recharge,* The Institution of Water Engineers and the Society for Water Treatment and Examination (1964),

5.4   T.P. Barber, 'Effluent treatment in processing industries', *Process Biochemistry* (September 1970).

5.5   T.P. Barber, 'Special filtration techniques', *Chemical and Process Engineering* (December 1970).

5.6   B. Surfleet, 'The electrochemical treatment of industrial effluents', *Electronics & Power* (November 1970).